iURBAN: Intelligent Urban Energy Tool

RIVER PUBLISHERS SERIES IN RENEWABLE ENERGY

Series Editor

ERIC JOHNSON
Atlantic Consulting
Switzerland

The "River Publishers Series in Renewable Energy" is a series of comprehensive academic and professional books which focus on theory and applications in renewable energy and sustainable energy solutions. The series will serve as a multi-disciplinary resource linking renewable energy with society. The book series fulfils the rapidly growing worldwide interest in energy solutions. It covers all fields of renewable energy and their possible applications will be addressed not only from a technical point of view, but also from economic, social, political, and financial aspect.

Books published in the series include research monographs, edited volumes, handbooks and textbooks. The books provide professionals, researchers, educators, and advanced students in the field with an invaluable insight into the latest research and developments.

Topics covered in the series include, but are by no means restricted to the following:

- Renewable energy
- Energy Solutions
- Energy storage
- Sustainability
- Green technology

For a list of other books in this series, visit www.riverpublishers.com

iURBAN: Intelligent Urban Energy Tool

Editors

Narcis Avellana

Sensing & Control, Spain

Alberto Fernandez

Sensing & Control, Spain

River Publishers

Published, sold and distributed by:
River Publishers
Alsbjergvej 10
9260 Gistrup
Denmark

River Publishers
Lange Geer 44
2611 PW Delft
The Netherlands

Tel.: +45369953197
www.riverpublishers.com

ISBN: 978-87-93519-10-7 (Hardback)
978-87-93519-09-1 (Ebook)

Contents

Preface

What do I need? How many cartons of milk should I buy? How many tomatoes do I need? Can I afford that? When we do the weekly shopping list, what we really do is to match the amount of food we expect to consume with the money we can spend for it. Can we apply the same simple model to manage our energy usage?

The iURBAN project aimed to apply such model for energy production and consumption in cities. It developed and applied in real pilots in two European cities in Rijeka, the second biggest city in Croatia, and in Plovdiv, the second biggest city in Bulgaria. iURBAN used the SMART urban decision support system (smartDSS) to save energy, to reduce consumption, to alleviate the local grid in high peak times, to help citizens to save money, and ultimately to empower them to make informed decision in their households.

The iURBAN project has been a collaborative effort of 9 organizations from 6 different EU member and associated states (Spain, Italy, Germany, UK, Bulgaria, and Croatia), including 5 representatives of the industry (4 SME and 1 large industry), 2 research organizations, and 2 local authorities.

The consortium was formed by three ICT organizations (Sensing and Control in Spain, Vitrociset in Italy, and Integrated Environmental Solutions in UK) providing the necessary expertise with substantial backgrounds in the field of building automation (building energy management systems), Internet of Things (IoT), sensors and actuators, decision support systems, artificial intelligence software, communication technologies and deployment, sustainability, energy efficiency measurement and verification protocols, and commercialization of such technologies and services. Moreover, in the consortium, there was the participation of two organizations specialized in social and business sciences in the field of energy management and production. From one side, Fraunhofer ISE is an organization with advanced expertise in research of renewable technologies, and the participating department, represented by psychologist of proven reputation in the energy application arena, specialized in social behavior, social awareness, and social acceptance of energy measures and energy technologies. Complementing this, the Institute of Computer

Science and Social Studies from University of Freiburghas provided the know-how on business models for energy consumption and production and the acceptance of such business models by the different stakeholders. Also, two energy companies have participated in the project, both linked with the two municipalities selected for the validation phase. Finally, local authorities from Rijeka and Plovdiv reinforced the involvement of cities and citizens. Both cities are highly active in local, national, and international energy efficiency projects and are dedicated to reaching 20-20-20 goal, covenant of mayors, green digital charter among others.

The iURBAN project was started in October 2013 and completed in September 2016. It was one of the highest rated Smart City projects awarded by Europe's 7th Framework Program and funded by the European Commission (EC) with a total budget of 5,6 million €, of which 3,8 million € was by the EC.

Acknowledgments

The editors would like to thank the European Commission for their support during the whole journey of the project of both the proposal and the implementation phase. They would also cordially thank all the authors, contributors of this book, the members of the Smart City Advisory Group, and all the people who participated and devoted their time and passion to make iURBAN project an exceptional piece of work.

The editorial team
Narcis Avellana
Alberto Fernandez
Sofia Aivalioti

List of Contributors

Aidan Melia, *Integrated Environmental Solutions, Glasgow, UK*

Alberto Fernandez, *Sensing & Control, Barcelona, Spain*

Energy Agency of Plovdiv, *Energy Agency of Plovdiv, Plovdiv, Bulgaria*

Fabrizio Lorenna, *Vitrociset, Roma, Italy*

Jens Strüker, *University of Freiburg, Freiburg im Breisgau, Germany*

Karwe Markus Alexander, *University of Freiburg, Freiburg, Germany*

Marco Forin, *Vitrociset, Roma, Italy*

Mike Oates, *Integrated Environmental Solutions, Glasgow, UK*

Narcis Avellana, *Sensing & Control, Barcelona, Spain*

Sergio Jurado, *Sensing & Control, Barcelona, Spain*

Sofia Aivalioti, *Sensing & Control, Barcelona, Spain*

Stefan Reichert, *University of Freiburg, Freiburg im Breisgau, Germany*

List of Figures

List of Tables

List of Abbreviations

AI	Artificial Intelligence
ANN	Artificial Neural Network
API	Application Programming Interface
BEMS	Building Energy Management Systems
BU	Business Unit
CART	Classification And Regression Trees
CDH	Central District Heating system
CDSS	Centralized Decision Support System
CDSS-GUI	Central Decision Support System Graphical User Interface
CHP	Combined Heat and Power
CO	Carbon Oxide
CO_2	Carbon Emission
DER	Distributed Energy Resources
DR	Demand Response
EC	European Commission
ESCO	Energy Service COmpany
EU	European Union
FIR	Fuzzy Inductive Reasoning
GHG	GreenHouse Gas
GSPS	General System Problem Solving
GUI	Graphical User Interface
HTTPS	Hypertext Transfer Protocol over SSL
ICT	Information and Communications Technology
IPMVP	International Performance Measurement and Verification Protocol
IS	Information Systems
iURBAN	Intelligent URBAn eNergy tool
KNN	K-Nearest Neighbours
KPI	Key Performance Indicators
kW	Kilowatt
LDSS	Local Decision Support System

LDSS-GUI	Local Decision Support System Graphical User Interface
MBUS	Meter-Bus
NN	Neural Networks
OGS	Open Geospatial Consortium
PET	Privacy-Enhancing Technology
PKI	Public-Key Infrastructure
PPDP	Privacy-Preserving Data Publishing
PrP	Privacy Proxy
PV	PhotoVoltaic
QID	Quasi IDentifier
RESTful	Systems conform to the constraints of REST
RF	Random Forest
ROI	Return On Investment
RSA	Algorithm for public-key cryptography
SCDB	Smart City DataBase
SCDB-CDSS	CDSS part of the Smart City DataBase
SCDB-LDSS	LDSS part of the Smart City DataBase
SCPA	Smart City Prediction Algorithms
SEC	Security framework
smartDSS	SMART urban Decision Support System
SMS	Smart Metering System
SOA	Service-Oriented Architecture
TET	Transparency-Enhancing Technology
TLS	Transport Layer Security
TTP	Trusted Third Party
VPP	Virtual Power Plant
WA	Weather Analytics
X.509 Certificate	SSL Certificates (PKI)
XML	Extensible Markup Language
ZKP	Zero Knowledge Proof

1

Introduction

Narcis Avellana and Sofia Aivalioti

Sensing & Control, Barcelona, Spain

Abstract

The energy continues to be produced by exhaustible and polluting fossil fuels endangering the health of citizens and of the surrounding ecosystems. People around the world are concentrating in big urban centres resulting in an even greater demand of energy stressing the existing supply systems and the environment. This book is about the course of the iURBAN project which developed and validated sustainable energy management systems applied in two European cities using novel ICT technologies.

Keywords: Energy management systems, ICT technology, Energy efficiency, Decision support system.

Today, roughly half of the world's population lives in urban areas, consuming two-thirds of total primary energy and generating over 70% of global energy-related carbon dioxide (CO_2) emissions. By 2030, it is estimated that around 60% of the world's population projected 8.2 billion people will be housed in urban areas. This means that residents who live and work within a city will consume around 75% of the world's annual energy demand. If most of this demand continues to be met by fossil fuels, then cities maintaining a business-as-usual approach will experience large increases in CO_2 emissions, greatly endangering the health of citizens and surrounding ecosystems. Fortunately, many European Union (EU) cities are setting examples to follow in their quest to become energy independent, such as the Swedish city of Växjö, which meets over 54% of its energy demand with local renewable energy and Copenhagen which covers 43% of its energy needs.

The popularity of singular household-powering solar panel systems has increased notably in Europe and worldwide with many buildings to generate a considerable proportion of energy enough to cover energy needs while reduce their carbon footprint. Technology has been a key driver for such investments and the global shift to alternatives for cleaner and renewable energy sources. Additionally, Information and Communications Technology (ICT) tools assist in the management aspects of the energy production and consumption. In particular, software development, smart meters, actuators, sensors, and user-friendly interfaces help managers, household owners, and tenants to create their own energy plans for exploiting optimally the energy generation and control consumption. ICT tools can increase awareness on energy issues and empower users in all energy related decisions. Likewise, individual energy plans can be combined with data from nearby power plants to create energy "shopping lists" for an entire city. The main output is creating more efficient urban system powered by new and interactive relationships between citizens, energy companies, and local government.

iURBAN project created an intelligent "brain" that helps households, businesses, and public buildings to make their energy shopping lists. The project developed and validated a software platform that integrates different ICT energy management systems (both hardware and software) in two pilot cities, providing useful data to a novel decision support system that makes available the necessary parameters for the generation and further operation of new business models. The business models contribute at a global level to efficiently manage and distribute the energy produced and consumed at a local level (city or neighborhood), incorporating behavioral aspects of the users into the software platform and in general prosumers.[1]

iURBAN project aimed to achieve the following objectives in different phases (Figure 1.1):

1. **Integration phase**: To integrate existing and new smart metering systems for water, heating, gas, and electricity. During this phase, a complete monitoring and synchronization between the metering devices with the *Smart City Database* (SCDB) has been carried out.
2. **Smart decision support system development and citizen empowerment phase**: To develop a SCDB network where all the energy-related data generated in the city has been stored and a decentralized energy decision support system that collects real-time or near real-time data, aggregates, analyses, and suggest actions of energy consumption and production. These tools support contractual demand control schemes,

[1]Consumers and producers.

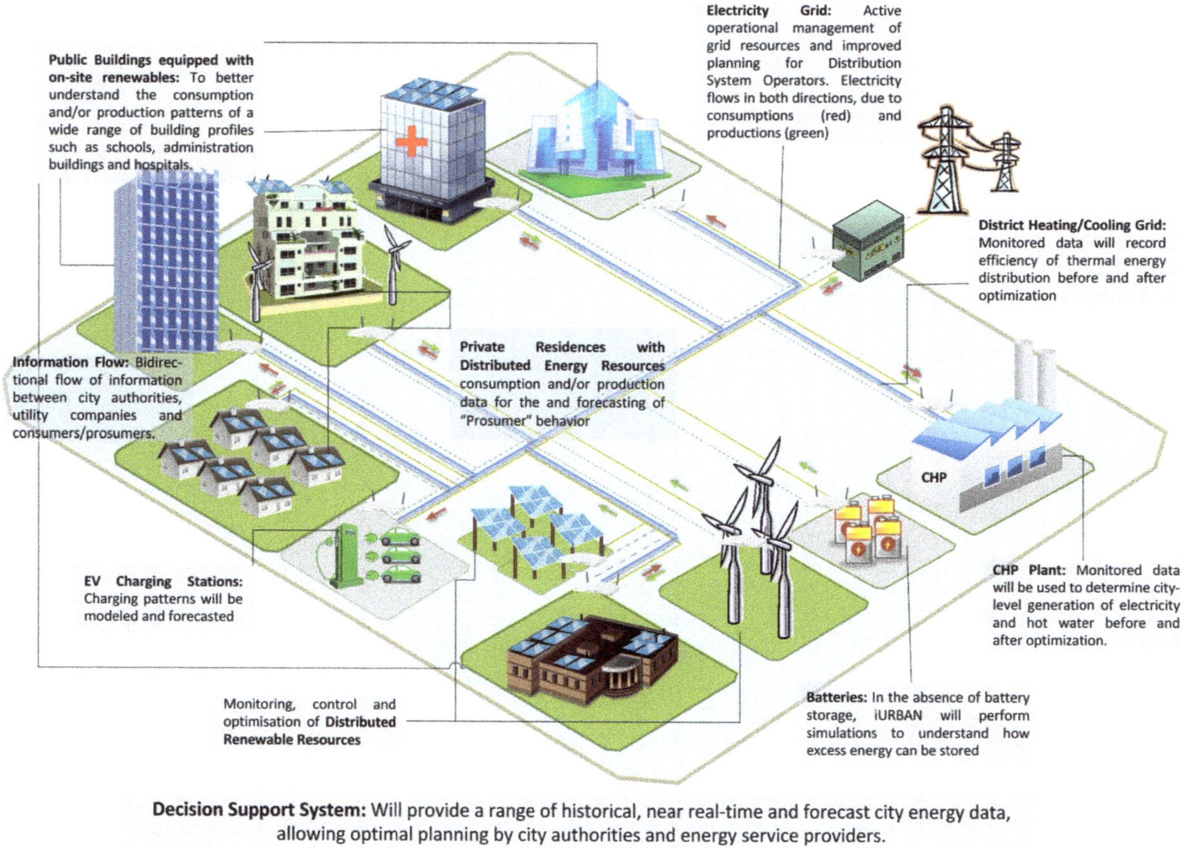

Figure 1.1 Overall scenario of the iURBAN project.

dynamic multi-tariffs, opportunities for load shifting, load shedding, storage utilization, prioritization of renewable energy, etc.
3. **ICT and socioeconomic evaluation phase**: To validate and evaluate the ICT tools together with the socioeconomic initiatives.

iURBAN project has conceived a SMART urban decision support system (smartDSS), i.e., a customized energy management and control platform in the framework of a city. The smartDSS adapts existing ICT and Building Energy Management Systems (BEMS) in a city and deploy it where necessary. The iURBAN smartDSS allows for scalability and incorporates a two-level decision support system:

1. *Local Decision Support System (LDSS).*
   LDSS engages consumers and prosumers by capturing near real-time data related to their energy consumption, as well as energy production from their installed distributed energy resources (DER), displaying it on a user-friendly interface via smart phones, tablets, PCs, etc., and provides support for decision making.
2. *Centralized Decision Support System (CDSS)*
   CDSS aggregates data from all LDSSs to provide city-level decision support to authorities and energy service providers. The *CDSS* generates a number of parameters, including citywide energy production and consumption forecasts. For example, the integrated utility in a city uses the CDSS which generates a forecasting for the next 24-hour consumptions and renewable productions in the different districts of the city, as well as advises for local energy market prices. Thus, the CDSS helps to balance citywide supply and demand with electricity generated by local renewable sources and requesting little energy from distant power plants.

   The LDSS communicates directly with the CDSS in order to exchange information for the decision making process. For instance, the LDSS receives the next-day energy prices and recommends to the end user that it is more cost efficient to wash her clothes between 10:00 am and 12:00 pm the next day.

The two-level *smartDSS* is a concrete solution comprised of the following:

1. A tool to measure, predict, and balance energy production, demand, and storage.
2. A tool to measure and verify reductions in energy consumption and greenhouse gas (GHG) emissions resulting from city energy use.

3. And user-friendly Web portals and functions that inspire new business models for all stakeholders (consumers, prosumers, city authorities, energy service providers, telecommunication companies, etc.). The decision support system leverages the cities' intelligent electrical and thermal grids to empower all of the actors involved and enable optimal distribution and trading of decentralized renewable energies production in a city, as well as the integration of combined heat and power (CHP) plants connected to the smart district heating and cooling grid.

The iURBAN *smartDSS* is an open software platform integrating local, intelligent energy systems and available communication infrastructures. Based on the nAssist[2] middleware, it serves as an interoperable data and information network, acting as a bridge between stakeholders. Interoperability with existing city infrastructure is ensured as well as taking into account new communication standards for electricity metering created by the International Electrotechnical Commission. *smartDSS* collects, aggregates, and analyses real-time or near real-time data from public and private city areas, integrating the ICT infrastructure which monitors and controls a wide range of energy-related devices found in European cities: low-voltage demand loads, distributed generation/storage resources,[3] smart district heating/cooling grids, weather data, smart meters, actuators, and electric vehicle charging stations. Outputs from the data analysis includes near real-time consumption and production data for each endpoint, from smart street lighting systems to a CHP plant to a commercial building, together with forecasts of demand response (DR) capabilities and consumption and DER production (Figure 1.2).

The *smartDSS* integrates all possible types of city endpoints. The endpoints in our two city pilots are found in various areas of the city and cannot be isolated on a single microgrid. Thus, DER units have been aggregated under the concept of a virtual power plant (VPP), where dispersed DERs have been integrated into one system. In addition, a simulation framework supports "virtual units" such as batteries, EV charging stations, and other distributed resources, inside the decision support system in order to analyze

[2]The nAssist© middleware platform is commercial property of Sensing and Control S.L. and is the basis of the iURBAN project adapted to accommodate city energy management, as well as the future integration of other city services such as security, health, transport, and waste management. http://www.sensingcontrol.com/es/.

[3]According to the IEC, electrical energy storage systems are considered to be a key enabler for wide-scale uptake of DER.

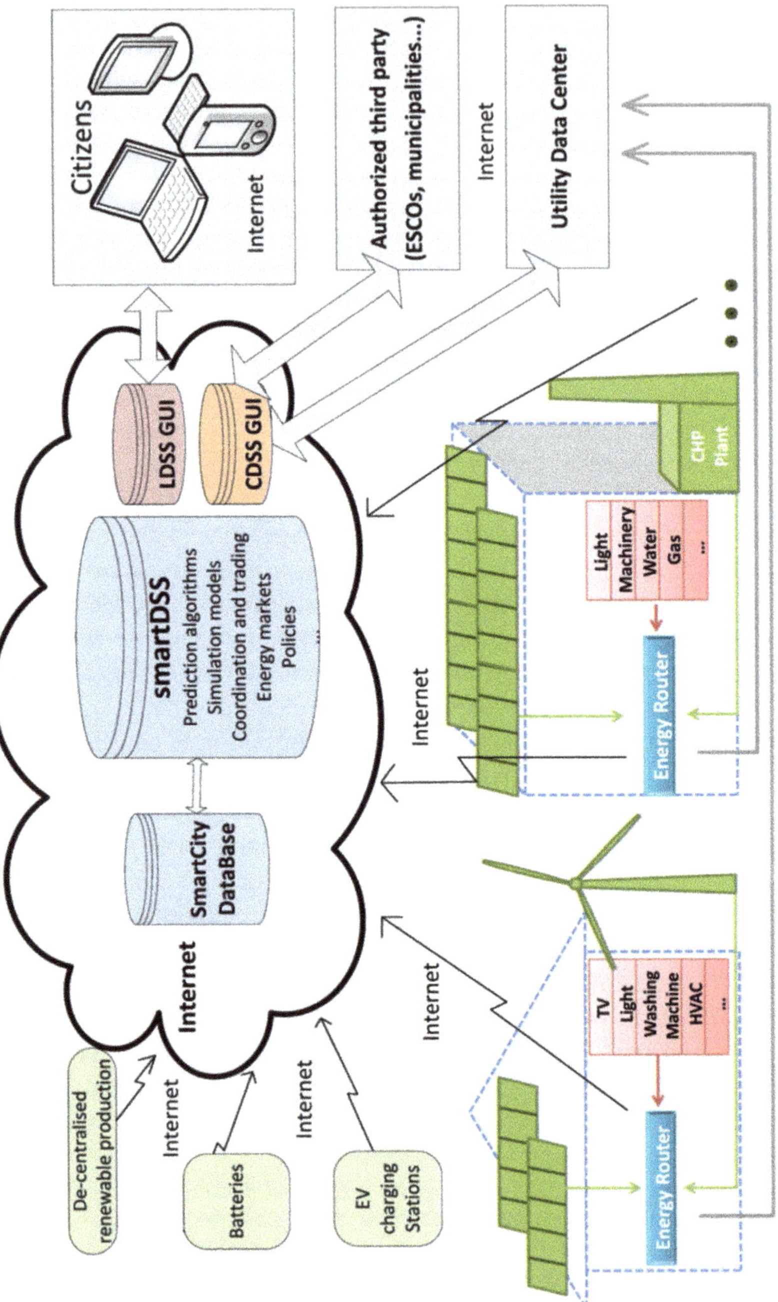

Figure 1.2 The iURBAN smartDSS architecture.

and understand how these virtual units could affect the current energy situation of the city in terms of consumption, production, CO_2 emissions, etc. These models are driven by external information and artificial intelligence (AI) models, and serve to complement the project's city pilot cases.

A key objective of the project was to develop and carry out a systematic procedure for the validation and evaluation of the impacts resulting from the deployment of the iURBAN smartDSS (Chapter 4).[4] Economic, environmental, and social key performance indicators (KPIs) have been drawn from the widely used International Performance Measurement and Verification Protocol (IPMVP)[5] and a social cost-benefit analysis has been performed, allowing for robust comparisons across heterogeneous European cities.[6] Moreover, this evaluation procedure aims to aid in the future exploitation of the system by demonstrating the system's return on investment (ROI) and its potential as a powerful political tool that city governments can utilize to justify energy policy decisions.

User engagement and openness are fundamental elements and serve as drivers for the design of the smartDSS. A major goal of the project was to enable city authorities to be able to provide to local actors, for instance energy agencies or energy service companies, with a wide range of data sets related to energy production, consumption, and infrastructure investments (i.e., the impact of investing in novel electricity storage units). Moreover, the project leverages knowledge from the behavioral science and behavioral economics fields that allowed the development of novel user-centric business models. Throughout the implementation of the project, surveys and focus groups have been carried out to identify and analyze all aspects that influence the acceptance of new business models in each pilot city, such as the actors' needs, cultural concerns, regulatory conditions, and data security issues. However, this book aims only on particular developments and focuses on the technological achievements of this 3-year work.

---

[4]smartDSS is an open software platform integrating local, intelligent energy systems and available communication infrastructures.

[5]"International Performance Measurement and Verification Protocol," International Performance Measurement & Verification Protocol Committee, 2002. http://www.nrel.gov/docs/fy02osti/31505.pdf.

[6]The cost-benefit analysis and key performance indicators included in this project are in accordance with EU Directive 2009/72/EC and are based on the methodology outlined the JRC Reference Report: *Guidelines for conducting a cost-benefit analysis of smart grid projects* (2012). The methodology includes indicators such as city-level economic impact of lower consumer utility bills, reduced healthcare costs due to less incidences of GHG-related respiratory problems, reduced congestion in the electrical grid and local job creation.

2

Logic Architecture, Components, and Functions

Alberto Fernandez

Sensing & Control, Barcelona, Spain

Abstract

This chapter provides an overview of iURBAN ICT, describing the overall architecture, main functionalities of each component and interfaces.

Keywords: ICT, Architecture, Functionality, Interfaces.

As shown in Figure 2.1, the principle idea behind the iURBAN platform and its architecture is to provide a set of outputs based on the inputs received by the smart metering infrastructure in Plovdiv and Rijeka pilots.

The architecture of iURBAN platform is mainly composed by three main blocks, as Figure 2.2 shows: the local decision support system (LDSS), centralized decision support system (CDSS), and the SMART urban decision support system (smartDSS). The three blocks communicate between them in order to provide a set of outputs; on one hand, the CDSS output is target to energy companies, municipalities, and other public authorities, while the LDSS is target to public and private customers. The smartDSS consists of a set of components that are used for both the LDSS and CDSS. Apart from that, the LDSS and CDSS have also its own components.

A global logic view of the architecture is depicted in Figure 2.3. The main components of iURBAN platform are the *Smart City Database* (SCDB), a set of open *interfaces/application program interfaces (APIs)* to connect third-party tools and services, and to provide support to internal modules of iURBAN. Within the *Orchestrator* is located the *smartDSS* and its business unit (BU), the LDSS, and CDSS. All of them under the security and privacy framework.

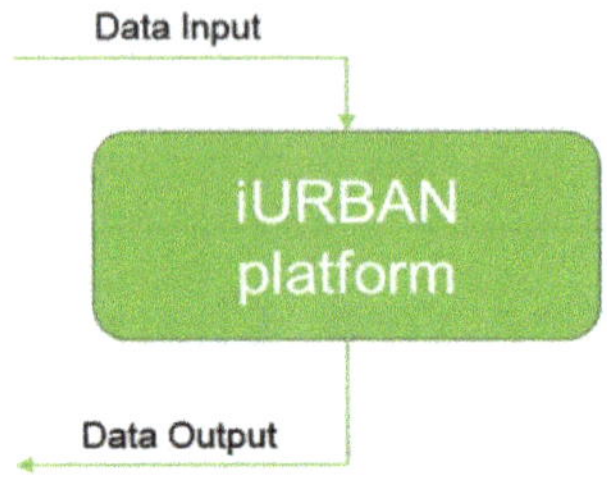

Figure 2.1 Data input–output iURBAN platform.

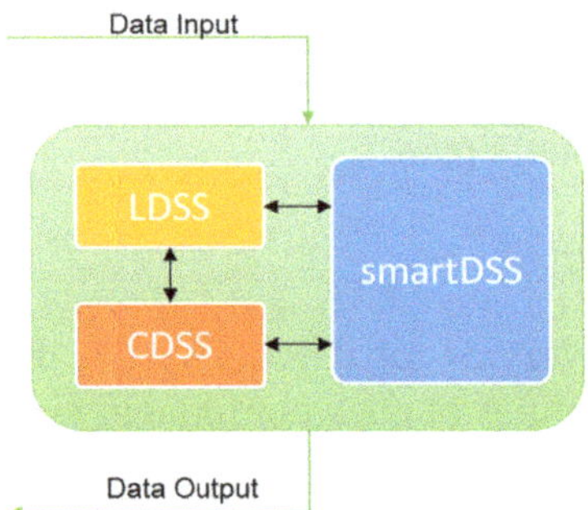

Figure 2.2 High-level logical architecture of the iURBAN platform.

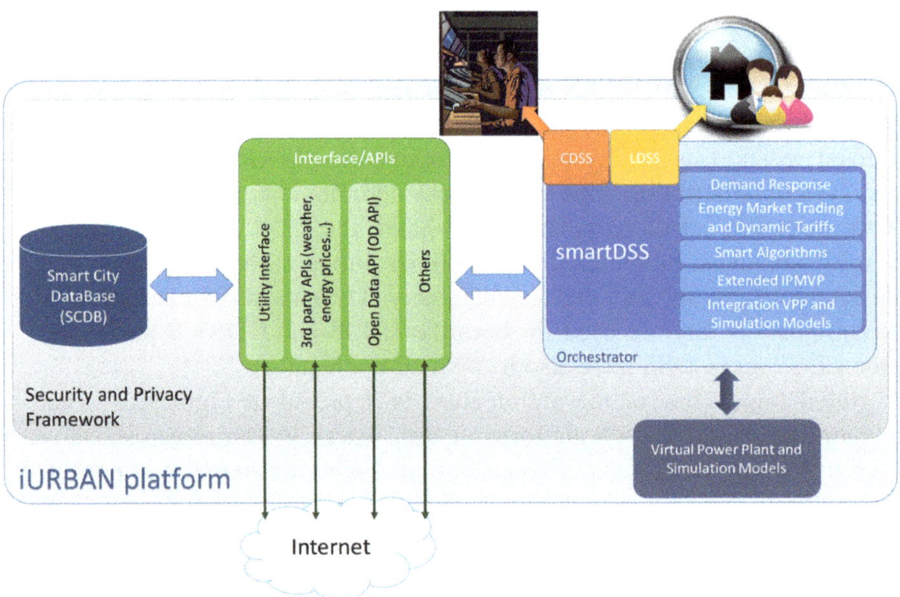

Figure 2.3 Logical architecture envisioned.

Energy consumed and produced and any related information with energy is provided by either internal or external services to iURBAN. All data is stored in the SCDB.

The four main services are as follows:

- **Utilities**: Service to receive energy consumption and production data from utilities' smart metering system.
- **Smart Home**: Smart Home provides contextual information such as internal temperature, humidity, CO_2 levels, and luminosity. The energy consumption can be obtained by smart plugs and sub-meters to obtain more detailed home consumption. Furthermore, it provides automation and security functionalities.
- **Energy Prices**: Service to receive energy prices, so the users can be up to date on bill expenditure.
- **Weather Provider**: Service to receive historical, current, and future weather information.

Figure 2.4 shows schematically all different function modules of iURBAN; furthermore, it relates each function to a main component (see figure's legend). Notice for instance that the *Information and Motivation* function module only affects the LDSS; thus, all components affecting this package are fully developed and integrated in the LDSS. Similarly, the *Technical Loses* function module only affects the CDSS; therefore, all components are inside CDSS package. Furthermore, the demand response (DR) function module affects LDSS and CDSS; then, a component to automatically identify potential DR actions is created within smartDSS components; in this particular case, additional components are also created in the LDSS (historical DR, notification, and advise) and CDSS (generation and validation) to support the use case.

2.1 Logic View

2.1.1 Local Decision Support System

The LDSS component of iURBAN is responsible to deliver a set of functions to households. It is composed by a back end and front end (Web- and mobile phone-based graphical user interface (GUI)) build on top of nAssist© IoT service from Sensing and Control (further information can be found in Section 2.2.1.4) (Figure 2.5).

LDSS main goal is to engage consumers and prosumers by capturing near real-time data related to their energy consumption, as well as energy production from their installed distributed energy resources (DER). The engagement

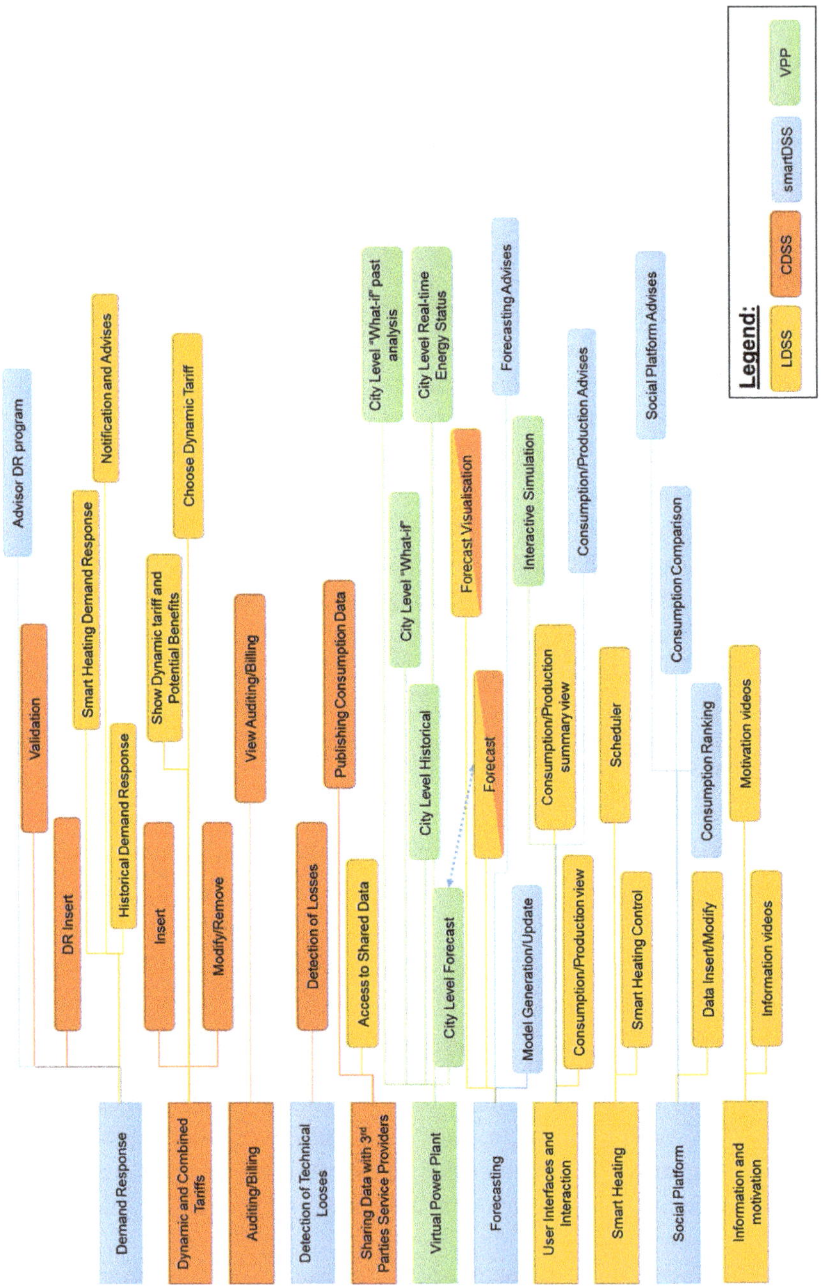

Figure 2.4 Business models UCs, user interaction UCs, and their impact with LDSS, CDSS, smartDSS, and VPP.

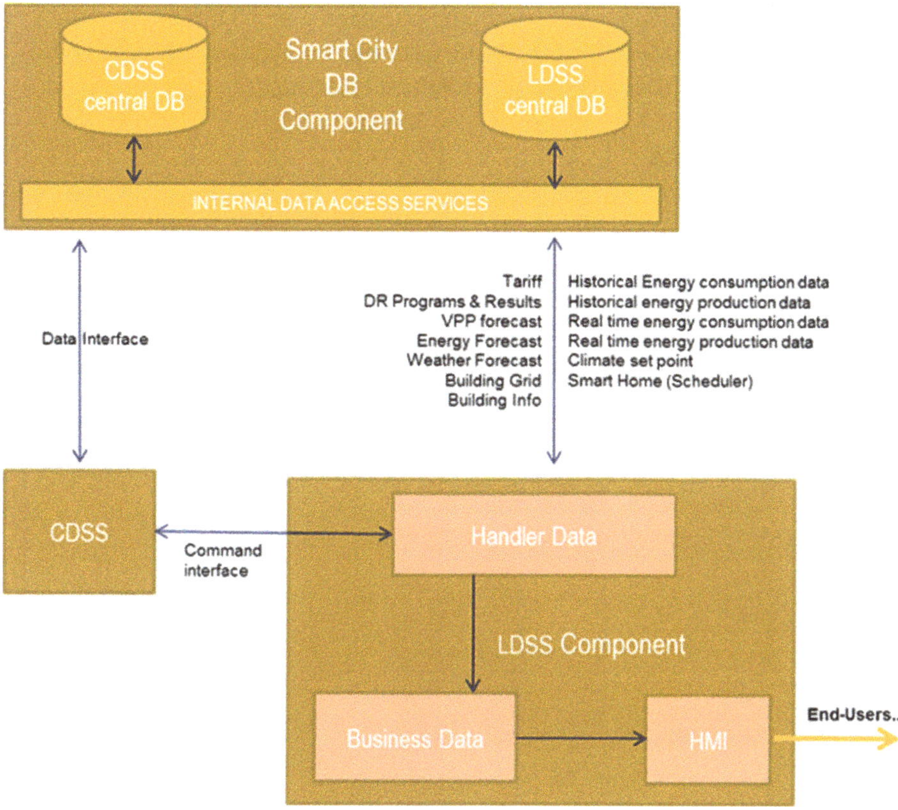

Figure 2.5 LDSS component and its relations.

is target throughout a user-friendly interface via smart phones, tablets, PCs, etc. Furthermore, it provides support for decision making; for instance, it advises how to achieve a DR actions, which dynamic tariff is more convenient, deviations compared to similar days/months/weeks and simulations to understand how different parameters affects to the user consumption and comfort. At this aim, the LDSS communicates directly with the CDSS in order to exchange information for the decision making process.

2.1.1.1 Handler data

The handler data is a sub-component of the LDSS component. It manages the interface from different components of iURBAN platform. This sub-component interfaces the LDSS component with the SCDB; it exposes two different interfaces: one for the CDSS central database and other for the LDSS local database to collect data to be sent to the business sub-component.

The handler data also manages command interface with CDSS component to send command in order to notify end user the availability of new DR program and the result of a given DR program ID executed.

2.1.1.2 Business data

The different sub-modules are depicted in Figure 2.6 and its functionalities are shown in Table 2.1.

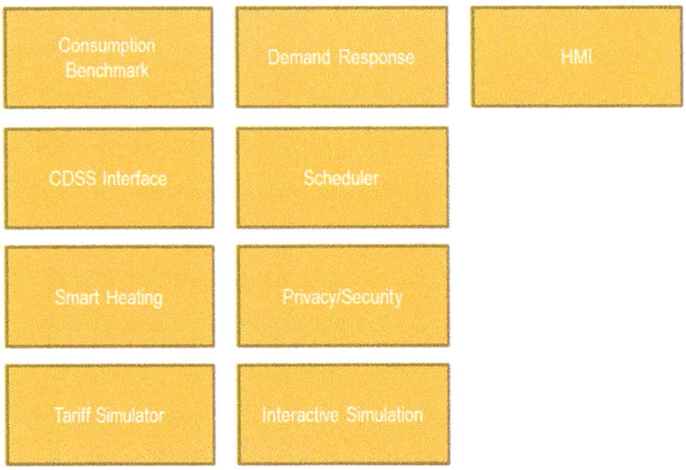

Figure 2.6 LDSS component.

Table 2.1 LDSS modules

Component	Description
Consumption benchmark	This module benchmarks your consumption and/or production against other peers.
Smart heating	This module implements the smart heating control of the home/business, plus the possibility to be controlled (target temperature) by the utility.
Tariff simulator	This module visualizes the different dynamic tariffs available and calculates the potential benefits of a time period taking into account its consumption profile (heating, gas, and electricity). It allows also to choose among the different available tariffs.
Demand response	This module implements DR programs, including notifications, historical DR actions, and smart heating DR.
Scheduler	This module implements the scheduler of DR program and smart thermostat control.
Interactive simulation	This module allows the users to see and simulate their consumption/production profiles in terms of kWh, CO_2 emissions, and EUR.
CDSS interface	This module provides a service interface for the communication between LDSS and CDSS.

2.1.1.3 Local decision support system user interface

This module provides the user interfaces of LDSS. The interfaces are developed on top of the current nAssist©, either modifying current ones or adding them to actual product. The new interfaces are tailored to increase user's satisfaction on iURBAN, and more concretely to support the user engagement activities and DR programs within the project.

Figure 2.7 shows the sub-modules of the HMI one, while further describes them. The default client and user interface platform is a Web browser.

The client implements those functions as visual components consuming data from LDSS RESTful API (using HTTPS connection) (Table 2.2).

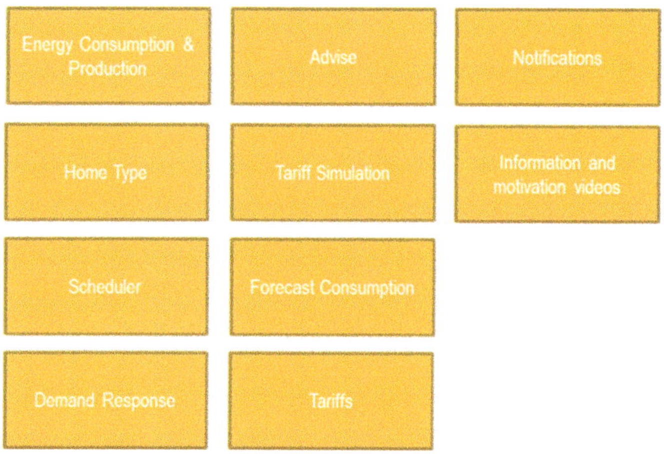

Figure 2.7 LDSS HMI interfaces.

Table 2.2 LDSS HMI sub-interfaces

Component	Description
Energy consumption and production	This interface provides to the end user a historical and almost real-time view of their consumptions and productions, provides both kWh and monetary cost. It differentiates between the different energy sources; electricity, water, and heating. The view is split in daily, weekly, and monthly information insights.
Home parameterization	This interface allows the user to insert home-related information that is used for the generation of personalized advice and the benchmark computation for energy consumption comparison. This information is additionally used for the user engagement.
Scheduler	This interface provides the scheduler for the smart heating use case. It allows to set a temperature in a concrete day of the week and hour (24 × 7 calendar).

(*Continued*)

Table 2.2 Continued

Component	Description
Demand response	This interface provides notifications about DR actions sent by the utilities and historical DR actions.
Advise	It provides advice for users with the aim to reduce their energy consumptions.
Notifications	This interface notifies the user about any communication with the CDSS.
Tariff simulation	This interface shows the available dynamic tariffs for the user, and it simulates its performance against the average consumption curve of this user. Moreover, it notifies when a more optimal tariff for the user is detected.
Energy forecast	This interface provides the (hourly, daily, 24/48/72 hours) energy consumption and production forecast. It shows an additional bar on plots with the real consumption and production.
Privacy policy	This interface provides to user means to decide the privacy policy configuration for his/her own data (energy and user information).
Interactive simulation	This interface provides a simulation of the energy building consumption performance. It allows to choose a set of parameters and verify how it affects to the consumption curve.
Home comparison	This interface provides the comparison between the user and its neighborhood.
Information and motivation videos	This interface provides a video or a link to a video with helpful hints, tips, and short easy videos on how to save energy. It also provides stories about peers telling why they use iURBAN platform and why is good, and what helped them save energy and money.

2.1.1.4 nAssist©

The core IoT platform of the LDSS architecture is the nAssist© IoT middleware platform, a commercial property of Sensing and Control S.L. It is provided to the iURBAN and adapted to accommodate city energy management, energy engagement functionalities, and advance energy service of the LDSS. The modular architecture of this platform allows it to be adapted very easily to other areas of application. The nAssist© platform has been developed following service-oriented architecture paradigm and therefore constituted by several tiers of processing. The module-based logic view is depicted in Figure 2.8. As shown, the software is divided into four main modules:

- Hardware/Software Interface: This module takes care of communication with field devices and external software services. The communication may be bidirectional depending on the type of devices connected and function to be used. Information transferred from remote sources is parsed, cleaned, formatted, and finally sent to upper software layer.

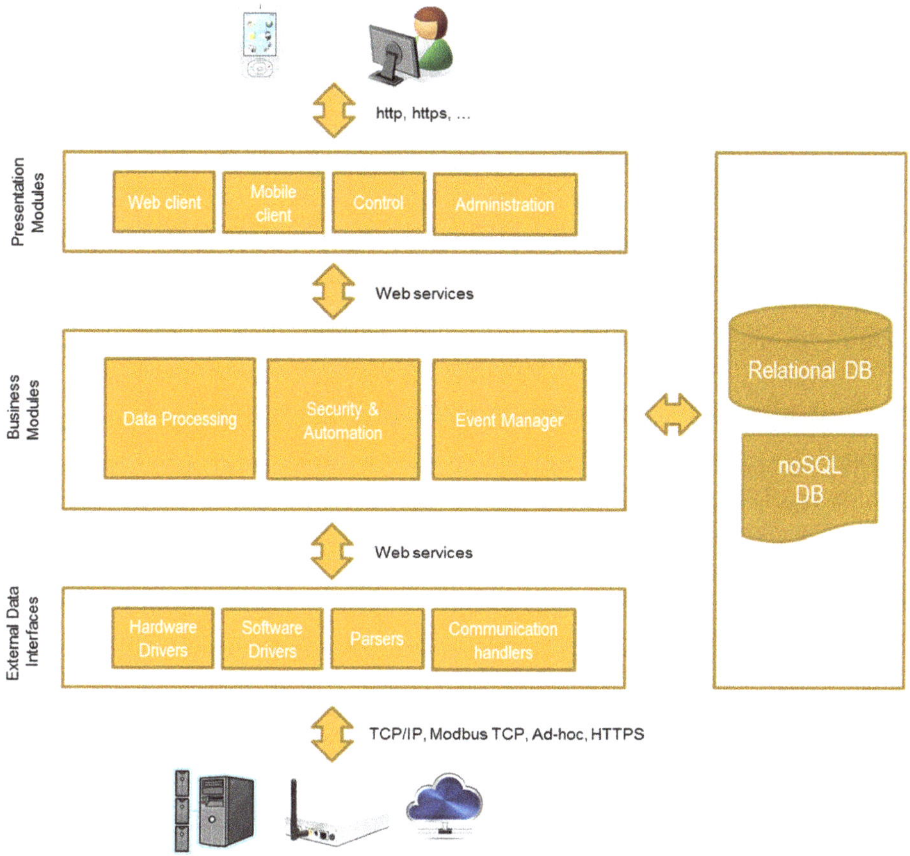

Figure 2.8 nAssist© logic architecture.

- Business Modules: The central module handles all the processing/intelligence of the software. It has a built-in event manager that handles communication between hardware interfaces and user interfaces, automation routines, and alarm/events handling and auto-generation based on rules. It handles database access.
- User Interfaces: The upper module of the software is the user interface, which works within standard Web clients and smart phones.
- Database: The core of the nAssist© platform provides a data warehouse and business intelligence components to analyze, manage, and comprehend data from devices and meter measurements. It has a relational database and a noSQL database.

nAssist© is built around special logic module called "framework foundations," which provides a common interface to access the low-level functions implemented by the middleware. This logic module works on two layers:

- Service layer: This includes business domains (user and application authentication, user and application registration, scheduling, etc.). Everything is accessible to both local and remote clients.
- Client layer: Clients are connected to service layer over the wire, meaning that those can work with server infrastructure if they have LAN or Internet connection. A client may be a Web site—Web-based products—or just a hardware sensor. Both end up using a RESTful Web service API.

2.1.2 Centralized Decision Support System

The CDSS component of iURBAN is responsible to aggregate and manage data at city level. That component delivers a set of functions to end users such as authorities, energy service company (ESCO), and municipalities, collecting data that come from other components of the system. The scope of CDSS is to integrate measurements data, with business models and energy production and consumption simulations.

The CDSS allows its users to perform the following activities:

- Get a continuous snapshot of city energy consumption and production.
- Manage energy consumption and production.
- Forecast energy consumption.
- Plan new energy "producers" for the future needs of the city.
- Visualize, analyze, and take decisions of all the end points that are consuming or producing energy in a city level, permitting them to forecast and planning renewable power generation available in the city, a real-time optimization and being perfectly scalable (meaning its ability to be enlarged to accommodate growth of data).

It composed by a back end and front end (Web- and mobile phone-based GUI). In this task, the GUI for CDSS is developed with a novel framework employing Web technologies (HTML5, CSS3, JScript, Websocket) to provide a Web 3.0 user interface experience. The different modules and sub-modules are depicted in Figure 2.9 and its functionalities are shown in Table 2.3.

The CDSS sub-components into the logical view are as follows (Figure 2.10 and Table 2.4):

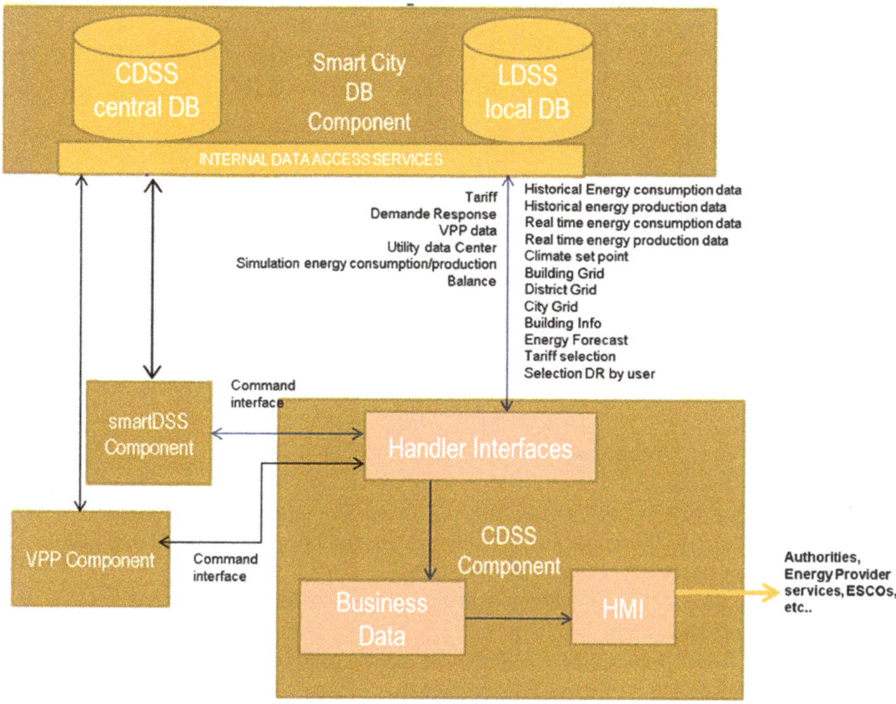

Figure 2.9 CDSS components.

Table 2.3 CDSS component

Component	Description
CDSS central database	It is the data warehouse database that also collects all the data at the city level. This database is used as data storage for business intelligent system.
Internal data access services	It is a sub-component that exposes services in order to communicate with other components internal to the platform as the CDSS, LDSS, VPP, and smartDSS.
Handler interfaces	It is a sub-component that has interfaces in order to • invoke VPP and smartDSS functionalities • invoke functionalities of internal data access services • invoke functionalities related to business data component
Business data	It is a sub-component that contains functionalities related to business intelligent system in order to perform historical data analysis.
HMI	This sub-component contains the CDSS graphical user interfaces for the end users.

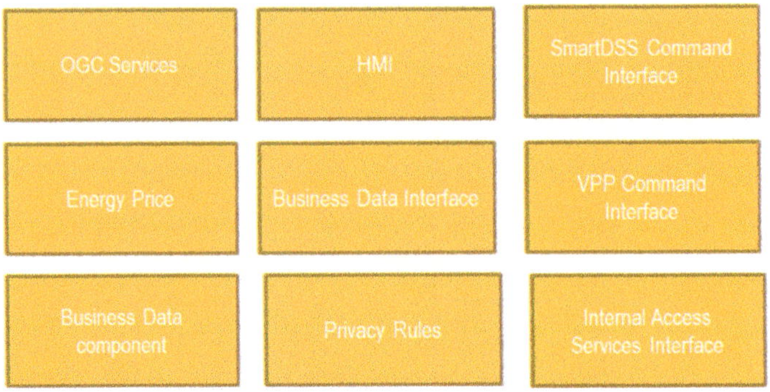

Figure 2.10 CDSS sub-components.

Table 2.4 CDSS sub-components

Component	Description
OGC services	It provides services in order to manage geographical information.
HMI	It is a sub-component to visualize data and information to the end user of the iURBAN platform side CDSS.
Energy price	It is a sub-component to manage tariffs information from utility provider price
Business data interface	This sub-component contains the interfaces to invoke functionalities related to business data component
smartDSS command interface	This sub-component contains the interfaces to invoke functionalities related to smartDSS component
Internal access service interface	This sub-component contains the interfaces to invoke functionalities related to internal access service sub-component
Privacy rules	This interface provides user means to decide the privacy policy configuration for his own data (energy and user information)
VPP command interface	This sub-component contains the interfaces to invoke functionalities related to VPP component

2.1.2.1 Centralized decision support system central database

The CDSS central database is a data warehouse database used for business intelligent system of iURBAN platform. This database is composed of two sub-components:

- Staging Area: It is a suitable database area that collects all the facts related to business intelligent system such as
  - Consumption (kWh)
  - Consumption (m^3)

- Production (kWh)
- Production (m^3).

These above data are read from LDSS database and stored to the staging area.

- Star Schema: It is a suitable database schema composed by fact tables and dimensional tables. Fact tables contain measurements related to consumption and productions and suitable foreign keys to dimensional tables. Dimensional tables contain information regarding data of
  - geographical
  - date–time
  - tariffs.

The goal of CDSS central database (that is a data warehouse database) is to have a storage that is used by suitable analysis functions in order to have historical data analysis.

This historical data analysis allows the user to aggregate facts (measurements) by dimension like:

- particular geographical zones;
- date–time intervals;
- tariffs intervals.

The tariffs are stored (to correspondent dimensional table of tariffs) from the energy utilities that provide energy prices. The tariffs are heterogeneous because each type of service operates on its own scheme; additionally, most measurements differ by nature (gas, heat, electricity). The smartDSS invokes functionalities of internal data access services in order to perform suitable data queries (from CDSS database) to obtain predictions regarding measurements (consumption or production).

The simulations data, coming from virtual power plant (VPP) component, are stored to LDSS database as described above. All the measurement data (simulated or not) are stored to staging area (of data warehouse) on the LDSS database. Figure 2.11 describes a logical view of CDSS component:

2.1.2.2 Handler interfaces

The handler interfaces is a sub-component into the CDSS component. It has suitable interfaces in order to invoke functionalities of different internal components of iURBAN platform. These interfaces invoke functionalities of components:

- VPP
- Business data

- smartDSS
- Internal data access services.

The handler interface also manages command interfaces. It sends commands to the following:

- VPP component in order to activate a processing of simulation to know how nonexisting "virtual" units can affect the consumption or saving CO_2.
- smartDSS component in order to perform prediction analysis of consumption and production (Figure 2.12 and Table 2.5).

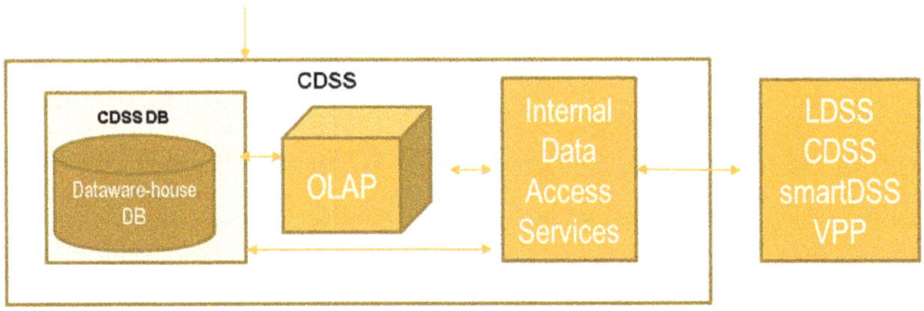

Figure 2.11 CDSS description.

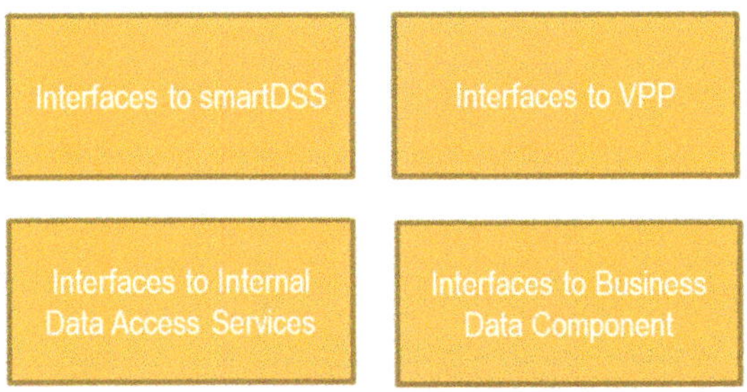

Figure 2.12 CDSS handler interfaces sub-component.

Table 2.5 Handler interfaces sub-components

Component	Description
Interfaces to smartDSS	It is able to invoke functionalities of smartDSS component
Interfaces to VPP	It is able to invoke functionalities of VPP component
Interfaces to internal data access services	It is able to invoke functionalities of internal data access services component
Interfaces to business data component	It is able to invoke functionalities of business data component

2.1.2.3 Business data

This is a sub-component that contains functionalities related to data analysis. In particular, these functionalities allow the user to

- Aggregate measurements of consumption/production following dimensions like
  - Date–time interval;
  - Geographical zones; and
  - Tariffs intervals.
- Detect technical losses: functionalities to calculate the differences between produced and consumed (by user) energy.
- Billing calculation (from consumed energy).
- Insert and validate actions related to DR.
- Dynamic and combined tariffs: manage the dynamic tariffs, update, and insert new dynamic tariff into SCDB (Figure 2.13 and Table 2.6).

2.1.2.4 Centralized decision support system HMI

The CDSS links and shows the data (including citywide energy production and consumption forecasts) generated from iURBAN with the end users via CDSS HMI.

The CDSS HMI component manages the interface with the authorities, the municipalities, the ESCOs, the utility provider services, iURBAN decision maker, etc., in order to display information as DR, dynamic tariffs, simulation VPP data, detection of technical losses, etc.

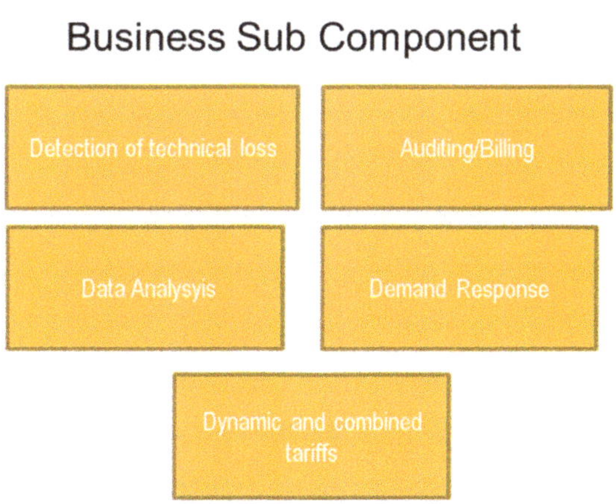

Figure 2.13 CDSS business component.

Table 2.6 CDSS business component

Component	Description
Detection of technical loss	To calculate the differences between produced and consumed (by user) energy.
Auditing/billing	Billing calculation
Data analysis	To aggregate measurements of consumption/production following dimensions like • Date–time interval; • Geographical zones; and • Tariffs intervals.
Demand response	To insert and validate demand response action and smart heating demand response
Dynamic and combined tariffs	To manage the different dynamic tariffs available. Inserts and updates or removes the dynamic tariffs from the system

The GUI allows monitoring the global loads of the grid, supporting the end user to plan suitable activities aimed at improving energy efficiency.

It also provides tools for visualizing and supporting the end user decisions concerning the best business models to apply. In case of existing building energy management systems (BEMS), the necessary adaptations for the correct functioning with the smartDSS should be developed in order to interface it to get data and be able to manage and control (Figure 2.14 and Table 2.7).

Manage communication channel is a sub-component to manage different devices and channels (mobile, portal, etc.) in order to communicate with end users (third parties, authorities, etc.).

iURBAN platform displays an overview of the grid status (electricity, heat and district hot water) for the level selected by the end user.

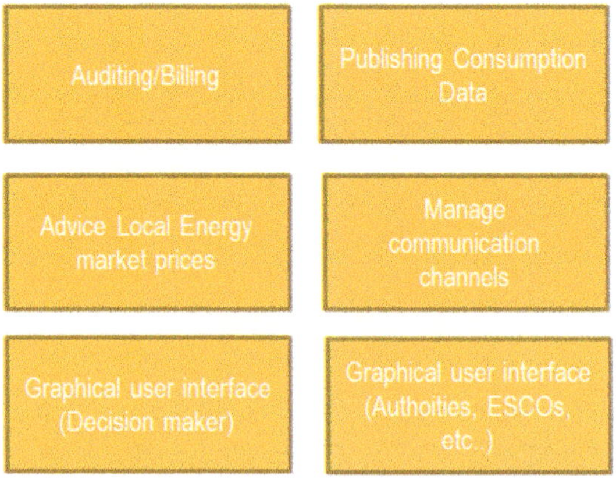

Figure 2.14 CDSS HMI sub-component.

Table 2.7 CDSS HMI sub-component

Component	Description
Auditing/billing	It is a sub-component to show auditing and/or billing information to the CDSS end user (third party, authorities, etc.)
Publishing consumption data	It is a sub-component to show publishing consumption data to the CDSS end user (third parties, authorities, etc.).
Advice local energy market prices	It is a sub-component to show results of predictor component in order to advice and help the CDSS end user (third party, authorities, etc.) taking the best solution about aspects of the market.
Manage communication channels	It is a sub-component to manage different devices and channels (mobile, portal, etc.) in order to communicate with end users (third parties, authorities, decision maker position).
GUI (decision maker)	It is a sub-component to provide an operative position for the decision maker. The info displayed is as follows: • Dynamic tariffs • Demand response • Detection of the technical losses • VPP
GUI (authorities, ESCOs, etc.)	It is a sub-component to provide a monitoring position for the authorities, ESCOs, and municipalities.

2.1.3 Smart Decision Support System

The smartDSS component of iURBAN is responsible to deliver a set of functions to be used by the LDSS and CDSS. It is composed by a back end with different business units (Figure 2.15).

smartDSS's main goal is to generate a set of insights on consumed and produced energy, based on the data stored in the SCDB. The principle idea behind the smartDSS is to generate valuable information and actions that cannot be taken by the LDSS and CDSS alone. There are common functions comparing LDSS, CDSS, and VPP; therefore, there is a need to have a common component capable to provide such output to the rest of components. For instance, the smart city prediction algorithms generate a set of outputs that are used by several packages or the DR actions that affects both the CDSS and LDSS. smartDSS is envisioned to group these components.

The different sub-components are depicted in Figure 2.16 and its functionalities are shown in Table 2.8.

2.1.4 Virtual Power Plant

The VPP component is responsible to deliver outputs stored on the SCDB, which in turn feeds as inputs into the CDSS. The relationship between the VPP and CDSS is as follows. The VPP is to act as the background calculation

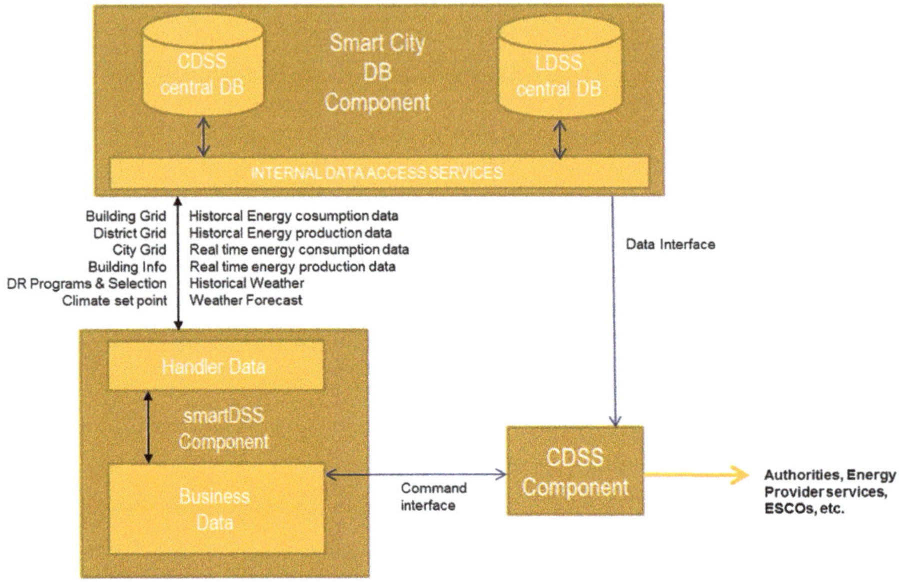

Figure 2.15 smartDSS components.

Figure 2.16 smartDSS sub-components.

engine to the CDSS, where the CDSS is to act as an interface to the VPP initiating user commands to the VPP. The different sub-modules are depicted in Figure 2.17 and its functionalities are shown in Table 2.9.

Table 2.8 smartDSS sub-components' description

Component	Description
Prediction algorithms	This module is responsible for the forecast models generation of the energy consumption and/or production for a given installation
Detection of technical looses	This module detects when there is a significant deviation from the smart meter data at building level and the sum of smart meters data at apartment level, belonging to this building
Consumption and production ranking	This module computes the different consumption and production rankings for each of the possible clusters.
Advisor DR program	This module advises to the CDSS about demand response actions that potentially could benefit to decrease load peaks and shift the load consumption curve. It generates also the optimal reward assigned to the demand response action.
Consumption/production advises	This module generates advises for the LDSS based on the historical and current energy consumption and production with the aim to reduce peak consumptions, CO_2 emissions, etc.
Social platform advises	This module generates advises for the LDSS to improve the energy ranking and contest.
Forecasting advises	This module generates advises for the LDSS based on the forecasted data with the aim to reduce peak consumptions, CO_2 emissions, etc.

Figure 2.17 VPP component.

2.1.5 Smart City Database

The SCDB is depicted in Figure 2.18. It contains the database of iURBAN and provides components to import data from external sources, as depicted in Figure 2.19, and additionally provides interfaces for accessing

Table 2.9 VPP modules

Component	Description
Energy metered data (building level)	VPP to access energy demand and generation metered per building/apartment from SCDB.
Smart algorithms (building level)	VPP to access smart algorithm data (demand, renewable, tariff) per building/apartment from SCDB.
Energy load aggregation (district level)	CDSS to send user commands to the VPP based on user defined and selected levels of interest in the model ranging from high-level city planning to the selection of individual buildings. VPP to carry out load aggregation of near real-time metered energy demand and generation data at building/apartment levels and data derived from smart algorithms, based on grouping schemes discussed above. VPP outputs to be stored on the SCDB for access by the CDSS and visually represented in the CDSS interface.
What-if simulation (building level)	CDSS to send user commands to the VPP defining what-if simulations to be carried out by the VPP calculation engine. What-if simulations are defined as simulation of additional distributed energy resources (DER), electricity storage, electric vehicles, etc., at user-defined building and/or city/district level. VPP outputs at building level to be stored on the SCDB.
Energy load aggregation (district level)	Same as above, except the focus here being on aggregation of near real-time metered energy demand and generation data at building/apartment levels, data derived from smart algorithms, and data from what-if simulations, based on grouping schemes discussed above. VPP outputs to be stored on the SCDB for access by the CDSS and visually represented in the CDSS interface.
Copy and paste metered data (building level)	User to identify lack of information in the CDSS city model (e.g., metered data only available for a small percentage of all buildings in the district/city). CDSS user option to copy and paste metered data at building level to buildings with insufficient metering within the city model.
Energy load aggregation (district level)	Same as above.
HMI	This module is led by CDSS, as the CDSS GUIs for the end users. VPP uses HMI as a user interface.

energy-related data collected from the main modules of iURBAN (LDSS, CDSS, smartDSS, and VPP). There is a special module, which allows access to data externally following the privacy and security policy of iURBAN (explained in Chapter 3).

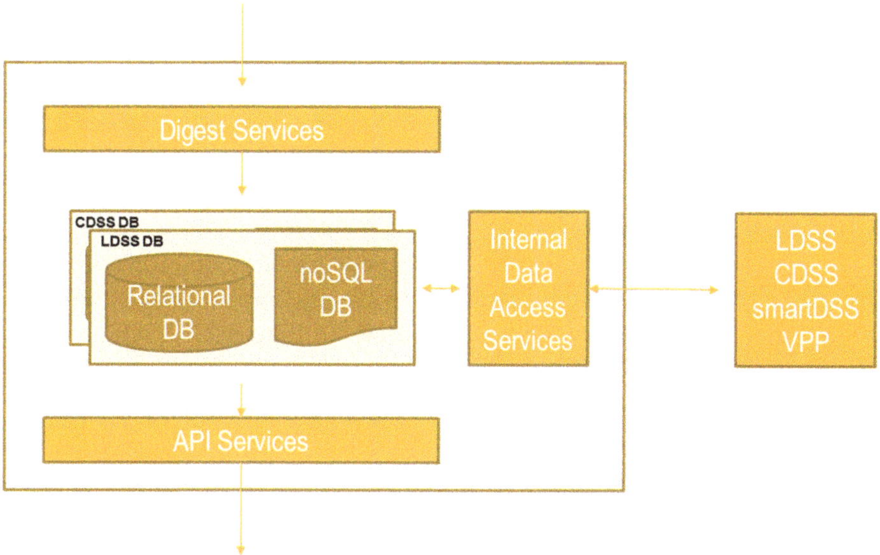

Figure 2.18 SCDB.

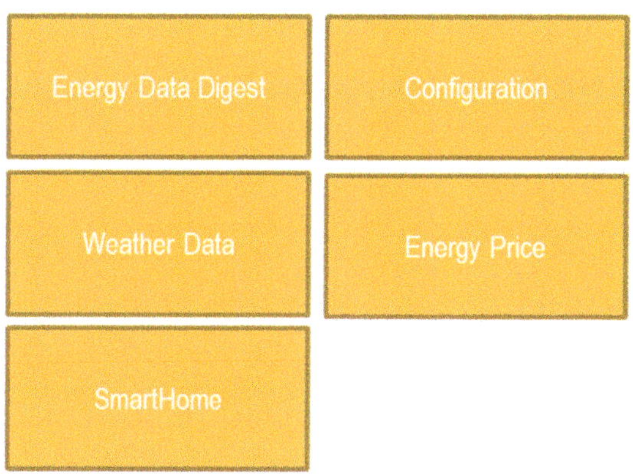

Figure 2.19 Digest component.

The database is divided into two main blocks, which matches the functionalities target for the LDSS and CDSS. Basically, all energy-related information from meter up to building level is stored in the LDSS database, while all energy-related information above building level is stored in the CDSS database; this include aggregated meter at city-level (district wise and city wise) data and VPP data. Both LDSS and CDSS incorporate a relational database, where the relation of the meters is defined: apartment/home/business, building, district, and city level, and a noSQL database to store massive energy information. Next sub-sections provide information about the different sub-components.

2.1.5.1 Digest component

The digest component of the SCDB is the responsible to acquire data from external sources necessary for business units to execute its function in correct way. It provides six different functions, as shown in Figure 2.19, which are further explained in Table 2.10.

2.1.5.2 Open data API services

The API service component is responsible for granting access to the energy information collected and computed within iURBAN following the iURBAN security and privacy policy. It provides an open API that allows third parties to offer external services to the users of iURBAN. This component provides a set of different as shown in Figure 2.20, which are further explained in Table 2.11.

2.1.5.3 Centralized decision support system database

CDSS database is a data warehouse database used by business intelligent system of iURBAN platform (Figure 2.21).

Table 2.10 Digest modules

Component	Description
Energy data digest	It provides Web services for the acquisition of meter data from the utilities. This component is responsible for storing meter data in the smart city database.
Configuration	It provides Web services for the setup of the smart city database relational content in relation with the utility meters.
Weather data	This module is responsible for fetching current and forecast weather data from external service.
Energy price	It provides Web services for the input of energy price.
Smart home	This module extends the current capabilities of nAssist© smart home interface providing new communication features in order to fulfill new smart home functions related to iURBAN user engagement and DR programs.

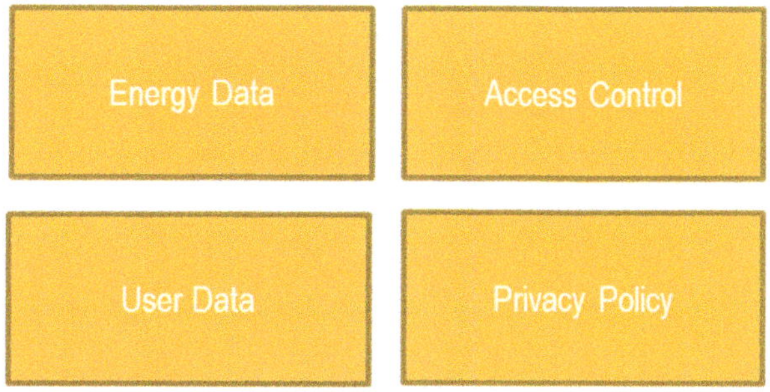

Figure 2.20 OPEN API module.

Table 2.11 OPEN API sub-modules

Component	Description
Energy data	This component exposes the energy data stored in iURBAN database. It provides access to current and historical data, from meter level up to city level.
User data	This component provides information from users following the privacy policy
Access control	This module controls which information can be access through the open API based on the request and the type of external service accessing the data. It logs all the request and all data access from external services.
Privacy policy	This module is used to set privacy policies for data access (explained in this chapter).

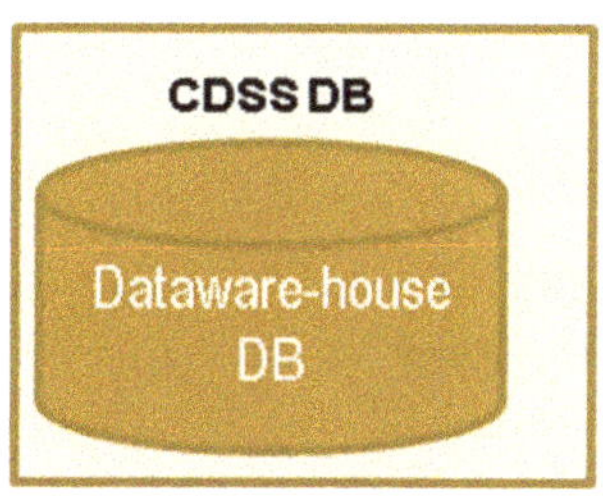

Figure 2.21 CDSS DB (data warehouse).

The data types stored to database are as follows:

- Consumption (KWh)
- Consumption (m^3)
- Production (KWh)
- Production (m^3).

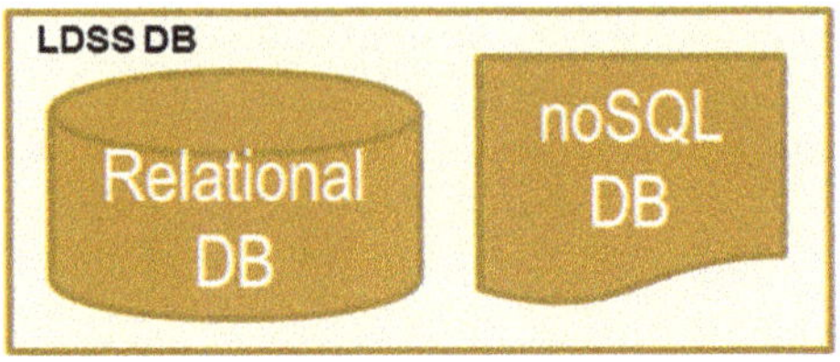

Figure 2.22 SQL/noSQL LDSS database.

2.1.5.4 LDSS database

The LDSS database (logic diagram in Figure 2.22) is built by adding new tables, relations, and storage on top of the nAssist©'s database.

The database is running in the cloud and it is only accessible through dedicated Web API provided by nAssist© enabled service.

Meter data is stored in the actual infrastructure of the nAssist© database. To support LDSS functionalities, the LDSS database stores additional information and relation as follows:

1. Grid infrastructure at building level.
    a. Electrical grid
    b. Heating grid
2. Tariffs
3. Demand response programs
    a. Historic events
4. Forecast calculations
    a. Hour basis
    b. Day basis
    c. 24/48/72-hour basis
5. Benchmark statistics at city level

2.2 Deployment View

The diagram below represents the typical production deployment of the iURBAN platform and can be viewed as a high-end configuration. Each sub-component of the platform is represented (Figure 2.23).

Almost all components of iURBAN platform are located online on the cloud in particular the LDSS sub-system. The cloud computing approach assures to the platform the following advantages:

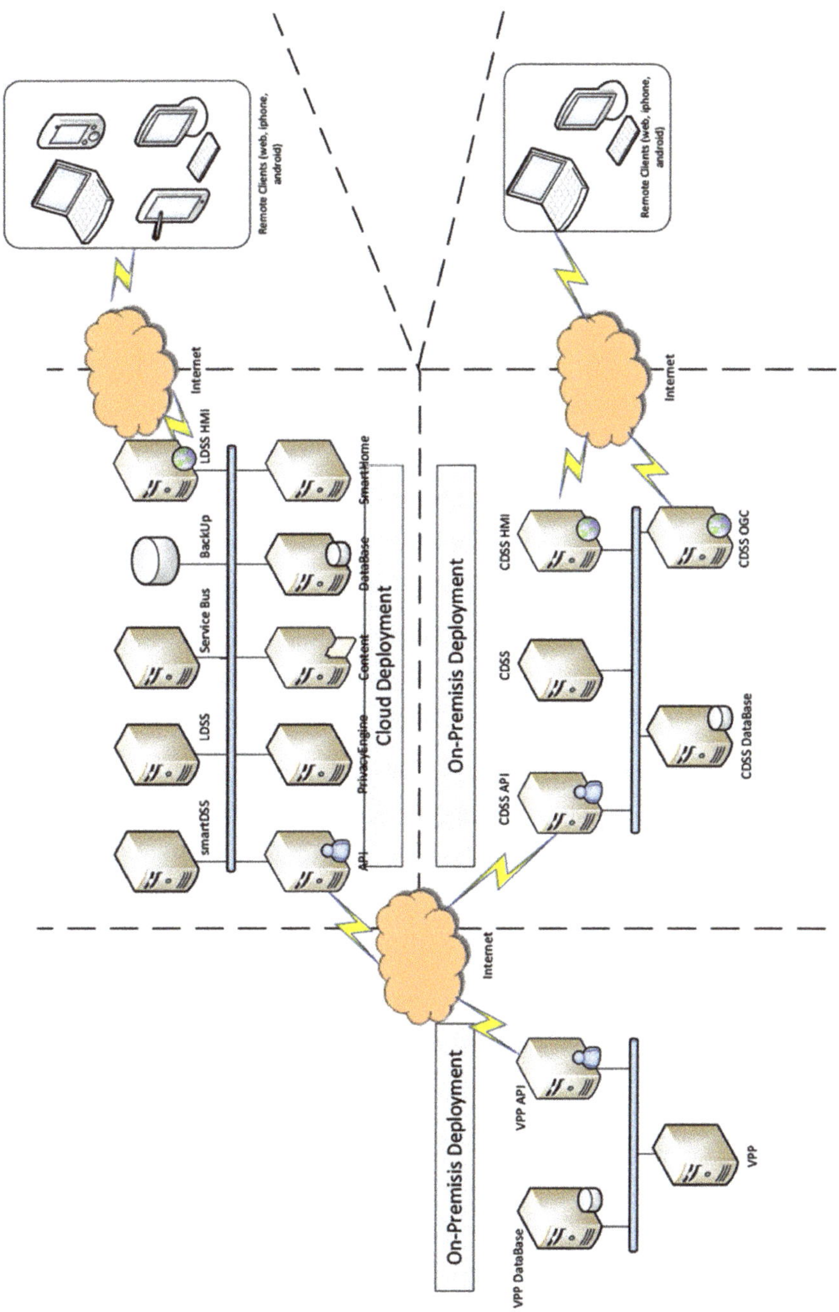

Figure 2.23 iURBAN deployment diagram.

- **Reduced Cost**: Cloud technology is paid incrementally, saving iURBAN consortium money.
- **Increased Storage**: The platform can store more data than on private computer systems.
- **Highly Automated**: No personnel need to worry about keeping software up to date.
- **Flexibility**: Cloud computing offers much more flexibility than other computing methods.
- **More Mobility**: Users can access information wherever they are, rather than having to remain at their desks.
- **Easy Deployment**: Centralized deployment.

However, in order to provide a high degree of data security and confidentiality, iURBAN platform is based on a "**Hybrid Mode**" deployment, which foresees to have critical components deployed on premises.

2.3 Conclusion

This chapter provides an overview about the main three ICT components of iURBAN tool and their relations. For each component, a sub-component explanation has been given, so the reader is able to have a wide overview of the functions and how those ones are mapped into its respective components and sub-components.

We have use hybrid architecture for deployment. Cloud infrastructure has been selected in order to account for main database, interfaces/API and intensive computation algorithms, and artificial intelligence models (predictions). This allows the reassign ICT and computational resources as smart grid meters grows over the time, which is expected to happen in parallel with number of residential users (LDSS users). CDSS and VPP have been designed to be deployed on premises, the ICT-demanded resources are lower, and it makes more manageable for local deployment.

3

Data Privacy and Confidentiality

Alberto Fernandez[1] and Karwe Markus Alexander[2]

[1]Sensing & Control, Barcelona, Spain
[2]University of Freiburg, Freiburg, Germany

Abstract

The transformation of the current electrical gird to a smart grid, enabling a real time analysis as well as response of electrical consumption, poses new security and privacy electricity grid challenges. It is of crucial interest for utilities to obtain precise consumption data, in order to manage the grid. From the security perspective confidentiality as well as integrity must be kept to ensure utilities receiving of correct data. From privacy perspective precise data poses a threat to customers. Precise energy data allows to gain a view into each participating household, which is beyond the original needs of performing grid management. The iUrban pilot builds a bridge between both contrary goals. Data needed for grid management is delivered in a precise form, while data for additional use cases, like analyzing energy consumption of a house, is delivered in a privacy preserving form.

Keywords: Security, Confidentiality, Smart Metering, Privacy, Privacy Enhancing Technologies.

Confidentiality (confidentiality, integrity, and availability)[1] is an information security requirement. We understand under confidentiality that only authorized entities shall be able to gain access to data. To ensure confidentiality, we consider the two concepts: access control and encryption.[2] Encryption means that

[1]http://security.blogoverflow.com/2012/08/confidentiality-integrity-availability-the-three-components-of-the-cia-triad/.

[2]Encryption can be seen also as a form of access control. Only entities which have the according key are able to decrypt properly encrypted data. Key management can be seen as access control.

cryptographic operations are performed on plaintext so that only authorized entities are able to decrypt it and read the retrieved plaintext. Access control means that only authorized entities are able to retrieve data from a system; in this case, the iURBAN platform. To ensure that only authorized entities are able to have access to iURBAN platform data, authentication is needed. To guarantee authentication, digital certificates as well as the distribution of log in credentials for platform users are performed. Three main concepts are used to protect the confidentiality of data in iURBAN:

- Security and encryption of the communication channels,
- Encryption of stored data,
- Access control to the iURBAN platform.

Figure 3.1 provides an overview of the iURBAN components connection. Between the components, between the data providers and the components, between the data providers and the users of the system as well as the smart city database (SCDB) component and the energy utility, the communication channel is protected in diverse ways. Data stored within the components is encrypted. Access to the system can only be obtained after an access control mechanism checks the validity of the request. The digital certificates as well as the access control rights need to be distributed under the platform users. The "iURBAN platform confidentiality manager" performs this work; this entity takes care for granting people and organization access rights to the platform, takes care for distributing the digital certificates, and is able to revoke both.

This is part of the administration area in iURBAN project. It is formed by a number of system administrators that have to be in charge of the system

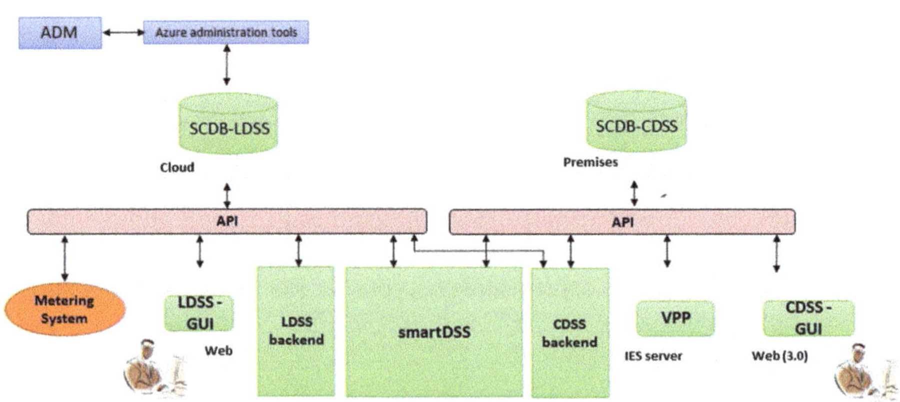

Figure 3.1 iURBAN components connection overview.

operational and security care. As in any other critical IT system, these technical personnel are in a trusted state by the company members of the consortium, but at the same time, they are not exempt of the security control to ensure all the operations they perform are conducted following the highest security levels. Lately, all the actions performed by these personnel are logged and securely stored for further review if necessary.

3.1 Confidentiality

The iURBAN system framework, as a whole, is seen as a service-oriented architecture (SOA) paradigm application and therefore constituted by several tiers of processing: data acquisition, data presentations, etc.

Data transmission networks transmit messages and commonly interconnect the several application tiers, which can include a mix of air and cable implementations. The end-to-end communication, across tiers, must be secured in order to ensure the basic security requirements of confidentiality, integrity, accessibility, availability, authenticity, and nonrepudiation and therefore protect the communication channels and messages that run across them (and even the ICT infrastructures).

The iURBAN security framework (SEC) is considering only Internet as the data transmission networks, because we are considering the utility networks as secure (Figure 3.2).

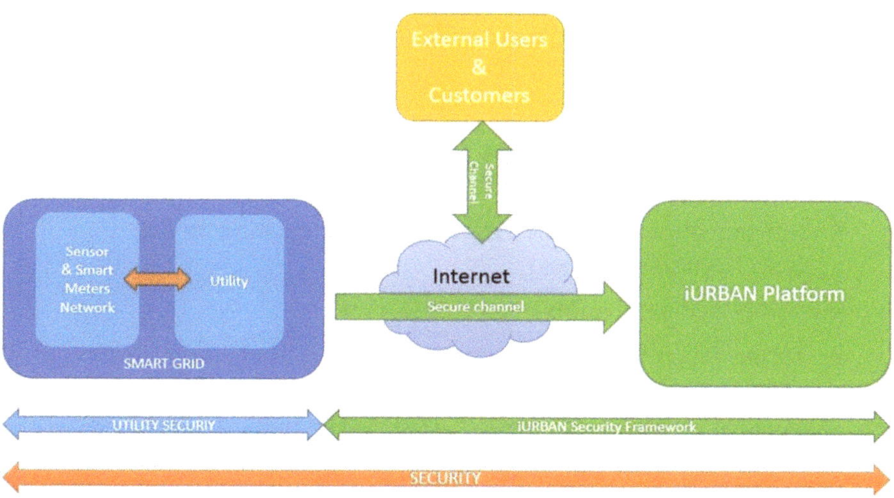

Figure 3.2 iURBAN security framework scope.

Confidentiality is an aspect of iURBAN nonfunctional requirements aimed at limiting information access and disclosure to authorized users and preventing access by or disclosure to unauthorized ones.

3.2 Confidentiality and General Security Requirements

To summarize, the security solutions implemented within the iURBAN framework in regard to confidentiality and other security-related aspects (EU directive 95/46/EC; EU recommendations 2012/148/EU; CEN-CENELEC-ETSI):

I. Confidentiality
   a. TLS/SSL protocol
   b. X.509 v3 certificates Class 1 with one-way authorization
   c. Azure traffic manager

II. Integrity
   a. TLS/SSL protocol
   b. X.509 v3 certificates Class 1 with one-way authorization

III. Authenticity
   a. MAC addresses control
   b. Authentication and authorization (the later existing on the administrators premises)
   c. TLS/SSL protocol
   d. X.509 v3 certificates Class 1 with one-way authorization

IV. Nonrepudiation
   a. TLS/SSL protocol
   b. X.509 v3 certificates Class 1 with one-way authorization
   c. MAC addresses control
   d. Authentication and authorization (the later existing on the administrators premises)

V. Accessibility and availability
   a. Offline data retention procedures
   b. High availability on components (load balancing/redundancy provided by Microsoft Azure Cloud)
   c. Backups/restore plans (provided by Microsoft Azure Cloud)

3.3 The iURBAN Privacy Challenge

The data minimization principle for smart metering data is in general implemented at the smart meter level. Such an implementation requires knowing in advance which question will be asked. The iURBAN platform is about to be able to ask questions about the data, which are not known in advance, creating the challenge to use a different approach.

Privacy preserving data publishing is a concept enabling to ask those questions, while preserving data privacy.

These privacy-preserving data publishing (PPDP) mechanisms are based on the interaction model where a data publisher uses collected data to issue them in a privacy-preserving manner (Figure 3.3).

PPDP mechanisms can be roughly classified into two categories based upon different attack scenarios and according to protection requirements. The concepts of the first one try to prevent that an attacker links records, attributes, or tables to a single person. The concepts in the second are based upon the uninformative principle: "*The published table should provide the adversary with little additional information beyond the background knowledge. In other words, there should not be a large difference between the prior and posterior beliefs*" [1]. Both categories allow to infer information about howl groups of participants but information gained upon an individual is limited which is achieved by reduced data accuracy [2].

While the business potential for fine granular consumption values, provided by smart meters is promising [3], those values pose the threat of privacy sphere invasion. In the Netherlands, this threat was sufficient to put a smart

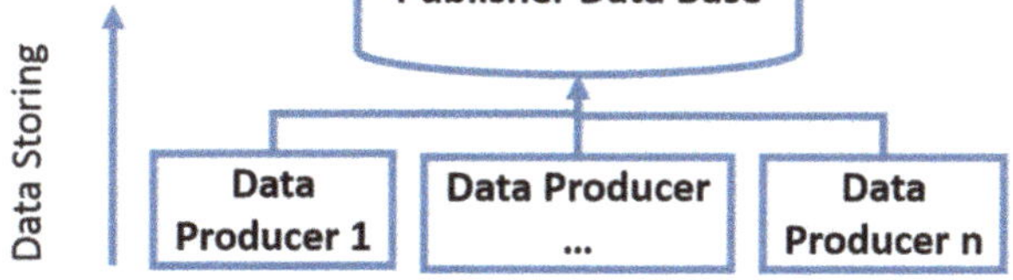

Figure 3.3 Privacy-preserving data publishing model.

meter roll out on hold [4]. By using customer data and energy values, a demand response aggregator (DRA) is able to find out customers' incentive and representing a considerable advantage on determining the compensation amount of an offer [5]. Computer science tries to avoid intermediaries and provides a variety of privacy-enhancing technologies (PETs) such as zero knowledge proofs (ZKP) where only one bit of information is divulged instead of a full consumption trace of fine granular energy data, creating the situation where the question be asked, must be known in advance [6]. The computing penalties are high and the approach is not promising in respect of managing smart grid.

Figure 3.4(A) shows a feedback loop of tasks for indirect load control (ILC). ILC is one option to create load shifts, and one form of it was implemented in iURBAN. Another option is direct load control, where a DRA can directly control devices at customer side or tariff-based programs where different prices at different times create incentives to shift load. Without an intermediary in ILC, a DRA makes a proposal directly to residential customers, e.g., households which accepts or rejects it. Out of the accepted, DRA selects a sufficiently large group to perform the intended load shift and sends this group the participation acknowledgement. For program verification and forecasts, DRA obtains energy data from the according household smart meter. We assume that the only way to obtain plain energy data is via access to the smart meter.

In price-based DR as shown in Figure 3.5, a DRA sends price signals to the respective smart meter to influence customer's behavior consumption. In an

Figure 3.4 Demand response data and interaction.

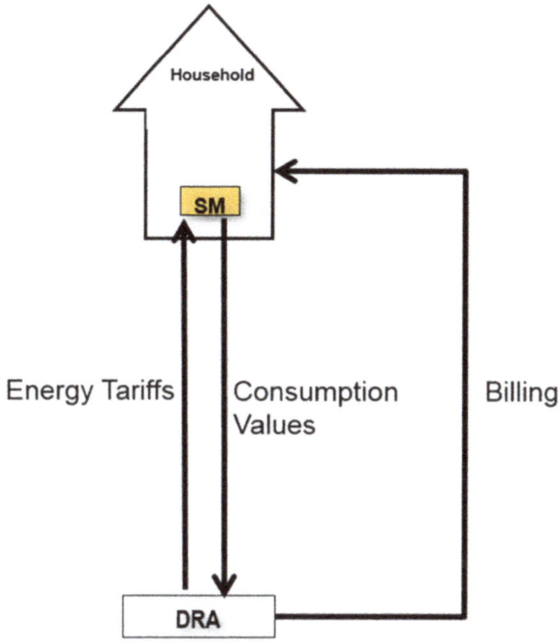

Figure 3.5 Price-based demand response and interaction.

intermediary setting, the signals are sent through the intermediary forwarding it to related recipients.

In both cases, the feedback loop allows DRA to exercise control of households. The loop needs data traces in order to be performed. Those traces leak privacy-sensitive information and thus are a potential privacy threat. Figure 3.4(B) shows how this feedback loop is interrupted by an intermediary. The iURBAN platform acts as the intermediary as shown in Figure 3.6.

In this feedback loop, all energy data are stored within the SCDB. The other components are accessing the data via the SCDB interfaces. This approach considers SCDB as a central cardinal point to protect PPDP. iURBAN follows the privacy by design approach which consists of seven principle steps:

- Proactive not reactive, preventative not remedial
- Privacy as the default setting
- Privacy embedded into the design
- Full functionality—positive sum, not zero sum
- End-to-end security—full life cycle protection

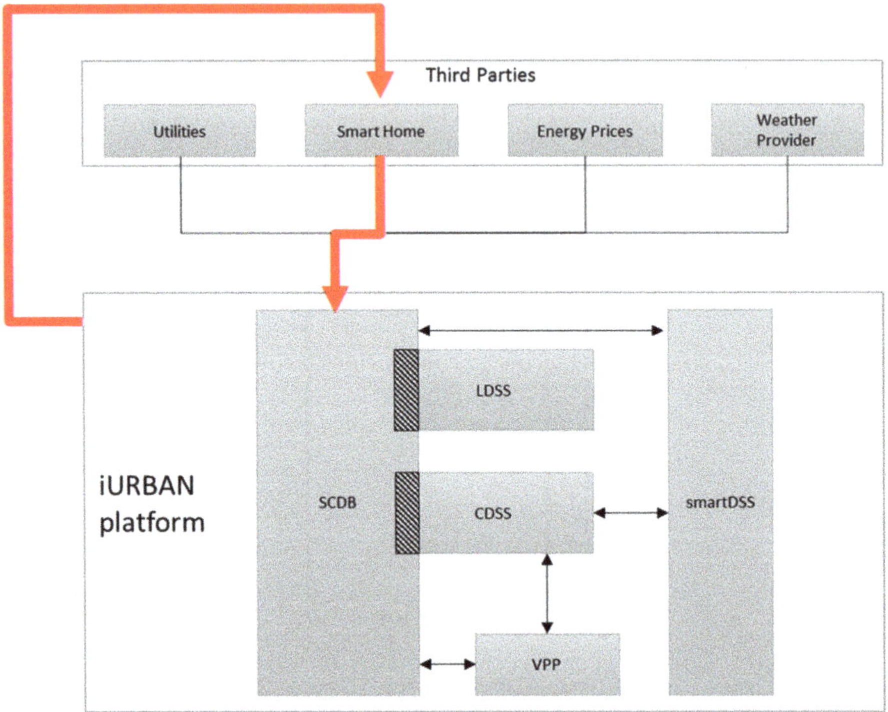

Figure 3.6 Logical architecture for iURBAN platform feedback loop.

- Visibility and transparency—keep it open
- Respect for user privacy—keep it user centric.

As intended by EU directive 95/46/EC, the privacy preservation concept of iURBAN is based upon PET and transparency enhancing. The fundament of these concepts is confidentiality as described in the first part of this section. The circumstance that the guarantee holds even in case of arbitrary background knowledge as well as the composability property makes it the suitable technique for the iURBAN project from privacy protection perspective. The impact on data utility and thus the impact on the energy goals need to be balanced with the required level of privacy protection. For transparency enhancing, we use the opt-in approach intended by the amendment of the data protection directive. An opt-in option is provided to the data producer where he can choose to allow direct access to his energy data without further privacy protection.

3.4 Privacy Enhancing via Transparency

Transparency-enhancing technologies (TETs) aim to inform the user how his data are stored, processed, and used for. The new proposal for data protection regulation demands the following points:

- Individuals shall fully understand how their data are handled.
- Explicit consent of subject to process their personal data which are achievable via TET.

The explicit consent is achieved by providing choice for different levels of data protection. There are several options to provide plaintext access to energy consumption data differing in the level of granularity. Another option is to provide consent for providing their data under the differential privacy guarantee. This choice determines the individual data protection policy. The collected data can be used for all iURBAN platform activities.

Each individual data producer is informed how his data are collected, stored, and processed via local decision support system (LDSS). It achieves this by providing information texts and graphics during the procedure to provide consent.

The user is able to revoke his consent. In this case, the past as well as future data may not be used.

3.5 Privacy Enhancing via Differential Privacy

Following the PPDP model shown in Figure 3.7, SCDB is the data publisher database within the iURBAN architecture. To ensure that data retrieval is only feasible in intended ways, we envision a privacy proxy regulating the access to SCDB and providing answers to request only in a privacy-preserving manner. This privacy proxy encapsulates the SCDB and acts as a trusted third party (TTP). While incoming data by legitimate users are passed to the database directly, all outgoing data are processed to keep privacy of users.

Role-based access control is applied to ensure that only legitimate data providers are allowed to put data into the database. Via interfaces and APIs, the privacy proxy receives queries. By access control, it is ensured that only legitimate users can successfully query. For each query, the privacy proxy (PrP) checks whether it is within the scope of the iURBAN privacy policy, respectively checks whether the budget is not exceeded or the request contains data producers who opted-in and those who did not. The proxy retrieves the data from SCDB and processes it to protect the privacy according to policy.

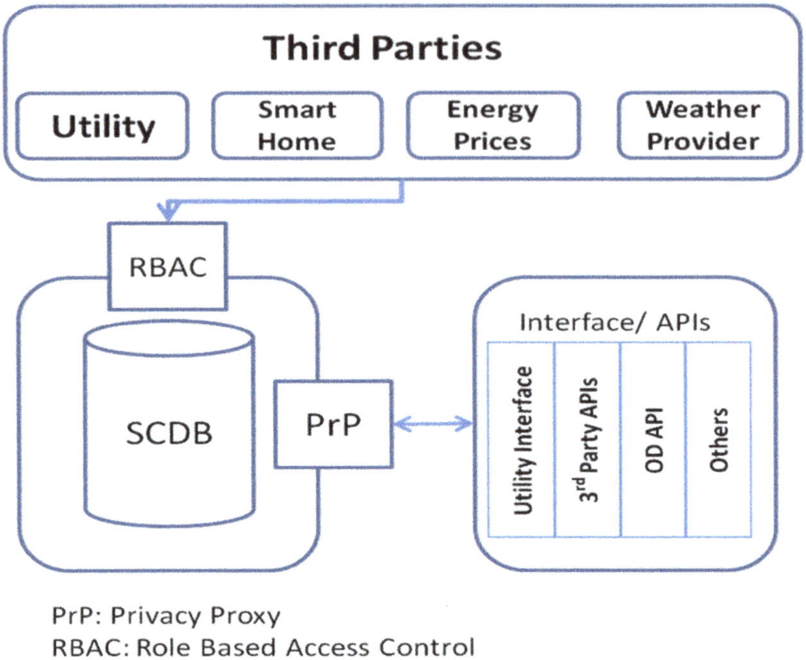

Figure 3.7 Logical view database privacy proxy.

Only this privacy protected answer is provided to the inquirer. Note that the only way to retrieve information from SCDB must be via the privacy proxy.

3.5.1 Privacy-Enhancing Technologies Based on Privacy Protection

The privacy proxy applies two kinds of privacy protection. The first one is temporal aggregation of consumption values of single data producers if they opt-in accordingly. Thus, if the customer stated he allows only access to energy data for a whole hour, the proxy sums up only complete four quarter hour measures.

The second option to protect privacy is to hide values of single customers within a privacy user group, if they did not grant access to plain energy values. This is achieved by using techniques providing the guarantee of differential privacy. Note that it is not possible to request data from producers who opt-in for plaintext access combined with data producers who did not under the differential privacy protection [7] (Figure 3.8).

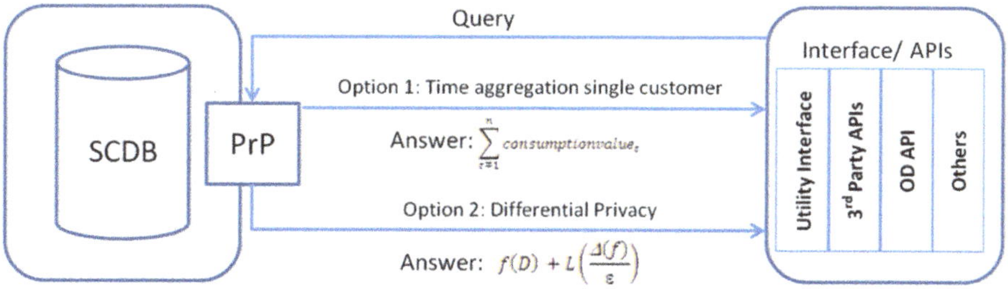

Figure 3.8 Logical view privacy protected query answers.

To mitigate the privacy threats for energy values, requests for differential privacy protected data shall be allowed as long as the following privacy requirements holds:

A user of the iURBAN platform shall not be able to learn an exact energy consumption value of a 15-minute interval for a single data producer.

To keep this requirement, a privacy budget for requests per data consumer ($\varepsilon -$ budget) must be settled. As soon as $\varepsilon -$ budget is used up, no further request shall be allowed. SCDB is continuously updated by the data producers. Differential privacy as a worst case guarantee associates this budget to the whole lifetime of the database with the consequence that in case of a used up budget, future requests of future energy values would be blocked. To prevent such a situation, $\varepsilon -$ budget is associated with time frames. Only requests for this frame are blocked, if the budget of the frame is used up. A suitable $\varepsilon -$ budget as well as an appropriate time frame needs to be evaluated during the course of the project.

3.5.2 Privacy Protection Implementation

The implementation of the privacy proxy approach is based upon the PINQ differential privacy framework [8]. It is written in C# and follows the privacy budget concept. The framework enforces differential privacy but does not define the access policy to the data base. For enforcing differential privacy, PINQ implements as aggregation operations NoisySum as well as NoisyCount achieved via Laplace-based noise mechanisms and NoisyMedian plus NoisyAvg achieved via an exponential noise mechanism.

The privacy proxy enforces that a differential privacy-preserved request does not contain data of customers who opted-in for providing their precise data. For enforcing access control, the privacy proxy relies on access control mechanisms of the underlying database.

3.6 Conclusions

This chapter has provided information about the approach implemented within iURBAN with respect to privacy and confidentiality of energy information being captured and stored from the smart grid.

As iURBAN stores public (public buildings) and private (households) energy information, its APIs and interfaces have been built to allow the data to flow transparently or biased depending on the level of privacy that the user would like to maintain.

A proxy has been designed and applied over that data captured by iURBAN for residential buildings. The proxy induces some perturbation on the original data in order to maintain privacy, which depends on the time span of the request and the type of query, meaning that depending on the target of the use of data queried, the corresponding use case can be or not achieved; this implies that this approach can jeopardize the possibility to launch services that can provide benefits the household (which holds the ownership of the data).

References

[1] Machanavajjhala, A., Gehrke, J., Kifer, D., and Venkitasubramaniam, M. (2006). l-diversity: Privacy beyond k-anonymity. In *Proceedings of the 22nd IEEE International Conference on Data Engineering (ICDE)*, 2006.

[2] Chen, R., Fung, B., Wang, K., and Yu, P. (2010). Privacy-preserving data publishing: A survey of recent developments. *ACM Comput. Surv. (CSUR)*, 42 (4), 2010.
Confidentiality, Integrity and availability:
http://security.blogoverflow.com/2012/08/confidentiality-integrity-availability-the-three-components-of-the-cia-triad/

[3] Albert, A., and Rajagopal, R. (2013). Smart meter driven segmentation: What your consumption says about you. *IEEE Trans. Power Syst.*, 28 (4), November 2013.

[4] AlAbdulkarim, L., and Lukszo, Z. (2011). Impact of privacy concerns on consumers' acceptance of smart metering in The Netherlands, international conference on networking, sensing and control, Delft Netherlands, 2011.

[5] Karwe, M, and Strüker, J. (2014). Privacy in residential demand side management applications. *Smart grid security: Second international workshop, SmartGridSec2014*, Munich Germany, 2014.

[6] Jawurek, M., Kerschbaum, F., and Orlandi, C. (2013). Zero-knowledge using garbled circuits: How to prove non-algebraic statements efficiently. *ACM Conference on Computer and Communications Security 2013*.

[7] Dwork, C., and Smith, A. (2009). Differential privacy for statistics: What we know and what we want to learn. *J. Privacy Confident. 2009*.

[8] McSherry, F. (2009). Privacy integrated queries an extensible platform for privacy-preserving data analysis. *SIGMOND* 2009.

4

iURBAN CDSS

Marco Forin and Fabrizio Lorenna

Vitrociset, Roma, Italy

Abstract

Central Decision Support System (CDSS) is the component of iURBAN platform dedicated to those stakeholders that are interested in the analysis of the data related to energy consumption and energy production, as energy providers, municipalities and third parties.

In this chapter, all the functionalities and the tools provided by CDSS will be discussed and presented to the reader, in order to give a complete overview of the capabilities of this component.

Keywords: CDSS, Energy management, Tariff, Chart, Map, Diagnostic, Tools, Demand response.

4.1 Introduction

Centralized decision support system (CDSS) aggregates data from all LDSSs to provide city-level decision support to authorities and energy service providers. The CDSS generates a number of parameters, including city-wide energy production and consumption forecasts. The CDSS component makes available, to different end users (as energy utilities, authorities, and municipalities), a large set of functionalities to monitor and plan the energy consumption/production.

Starting from the analysis of the data context and the users' needs, we have defined and describe a detailed set of functional requirements that are implemented by the CDSS platform:

- Get a continuous snapshot of city energy consumption and production;
- Provide a near real time regarding the situation of consumption/production of the city for every type of energy (electricity, heating, gas, hot water and cold water);
- Manage energy consumption and production data;
- CDSS HMI: the granularity is the building. The user, of CDSS HMI, can filter aggregated measures by city, district, building, energy type, and time period;
- Forecast of energy consumption data and per meter as well;
- Plan of new energy "producers" for the future needs of the city;
- Provide what-if scenario feature, offered by CDSS HMI (in collaboration with CDSS back end and virtual power plant (VPP));
- Support static and dynamic tariffs;
- Manage demand response (DR) programs;
- Manage technical losses;
- Visualize historical data with a chosen granularity and time period of consumption and production of energy and cold water.
- Detect meters in offline status;
- Provide geographical maps of building locations;
- Analyze International Performance Measurement and Verification Protocol (IPMVP) and key performance indicators (KPI) metrics;
- Make available weather forecasts.

In Figure 4.1, a general architecture of the whole CDSS is described, including the connections (regarding data exchanges) with the other iURBAN components:

As depicted in the above architecture schema, CDSS is composed of some sub-components.

In the following, the flow of the data managed by CDSS is described:

- Energy providers store real-time and historical measures to LDSS database (a component of Smart City Database (SCDB)) by suitable local decision support system (LDSS) Web services.
- CDSS (the processes of CDSS back end) retrieves the historical and real-time measures, by suitable LDSS Web services, from the LDSS database.
- CDSS GUI (the front end of CDSS) exchanges data with CDSS DB (by suitable CDSS Web services) in order to display measures in the graphical user interface (GUI) or to store data (for example, DR programs) to CDSS database.

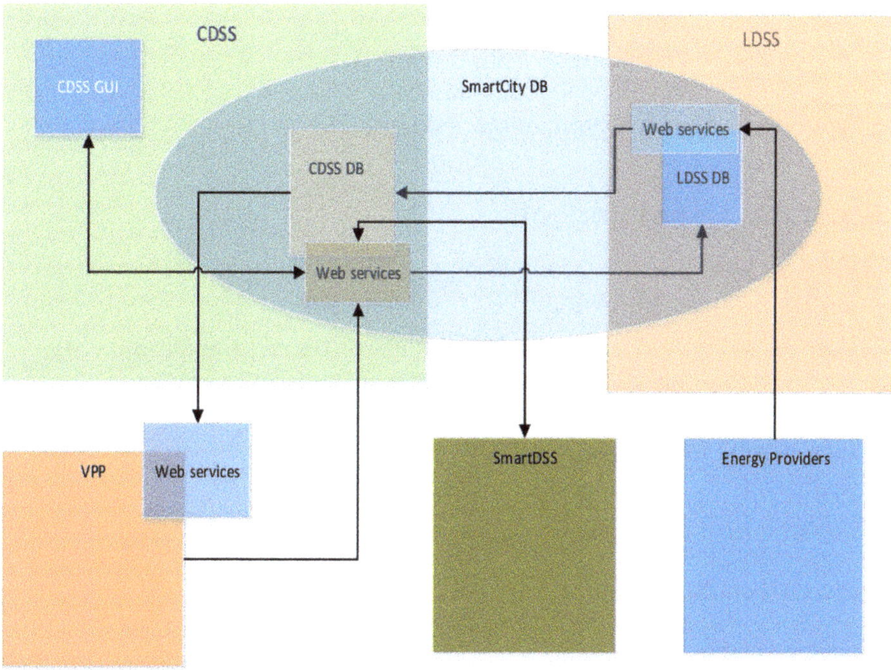

Figure 4.1 iURBAN logical view.

- CDSS exchanges data with the VPP in order to
    - VPP retrieves (from CDSS database by suitable Web services) the list of calculation (to do) related to what-if scenario.
    - VPP stores to CDSS database (by suitable Web services) the output result of calculation regarding the what-if scenarios.
- smartDSS exchanges data with CDSS in order to store data related to the following:
    - Calculated technical losses
    - Prediction algorithms
    - Advisor DR program
    - Consumption and production advice
    - Forecasting advice
    - Consumption and production rankings.
- smartDSS exchanges data with CDSS in order to retrieve data, from CDSS database, to calculate the following:

- Technical losses
- Advisor DR program
- Consumption and production advice
- Consumption and production rankings.

4.2 Graphical User Interface

The CDSS user interfaces are grouped by a suitable GUI that offers a set of functionalities to do the following:

- Check consumption and production with different granularity (city, district, building, installation) and filters that can be customized;
- Manage DR program management;
- Manage tariffs management;
- Visualize city map with involved buildings;
- Provide weather forecast;
- Offer meters diagnostic;
- Make GUI configuration.

4.3 Main GUI Functionalities in Detail

The main user functionalities, of the CDSS GUI, including the user authentication and the user roles (the user role indicates that the specified user can access only to a subset of user functionalities) are presented below.

4.3.1 User Login

The end users of the CDSS GUI are energy providers, local authorities, and municipalities but the login to the iURBAN system CDSS GUI is allowed to the following actors:

- **Super User**: They can view, create, and delete all data for every city and every energy type, but they cannot operate on CDSS DB data.
- **System Administrator**: They can view all data for every city and every energy type and can operate on CDSS DB data.
- **Decision Maker**: They can view all data for only one city and one energy type at all levels and cannot operate on CDSS DB data.
- **Provider**: They can view all data for only one city and only one energy type at all levels and cannot operate on CDSS DB data.

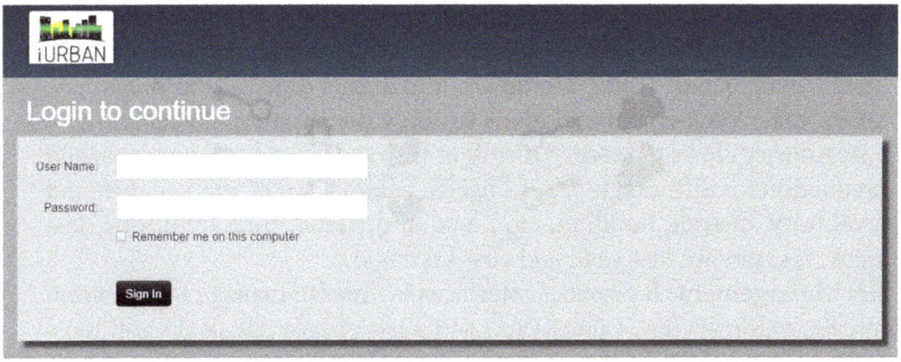

Figure 4.2 User login.

- **Municipalities, Local Authorities, Industry, and Similar Categories**: They can view all data for only one city and every energy type until city level; they cannot operate on CDSS DB data.
- **General User**: They can view all data for only one city and one energy type at city level and cannot operate on CDSS DB data.

The CDSS system manages the user profile and the authentication of the actors of iURBAN system CDSS GUI. The operator can interact with the system only through the features allowed by the assigned profile. The user reaches the GUI toolbar after an authentication consisting of the insertion of the user name and password in a dedicated pop-up presented before the access to the CDSS GUI (Figure 4.2).

4.3.2 Toolbar

Toolbar collects the macro functionalities of iURBAN interfaces. They are grouped taking into account the different kind of users (provider of energy, local authority and municipality, decision maker, etc.). The aim is to provide an easy and useful interface to allow the operator to manage data and information in an optimal way. From the toolbar, it is possible to open all the windows and the pop-up at the same time; this characteristic enables to view all data and to do comparison, analysis, and operations without information losses. The toolbar shows different views for each user profile, so the operator can interact with the system only through the features allowed by the assigned profile. The macro functionalities accessible from the toolbar are as follows:

- **Management**: It provides a cartographical representation of data and it shows a map of the city in 2D with districts, buildings, installations, meters using smart icons. Moreover, it manages objects (buildings, EV, DERs, etc.) like geometrical items to allow the creation of scenarios.
- **CityEnergyView**: It provides different views of energy consumption and production for different type (electricity, heating, water, etc.), at different level (city, district, building, etc.), and at different time (real time, last week, last month, last year, and custom range).
- **DR Management**: It provides interfaces in order to manage DR program (create, modify, view, cancel, etc.) and a list of possible peaks and tools to manage notifications to customers.
- **Tariff**: It provides interfaces for the tariff management and comparison.
- **Diagnostic**: It provides information about the status of the installations, whether they are offline and whether there are losses.
- **Weather Forecast**: It provides weather forecast information.
- **User**: It provides information about the current user login and manages its logout.
- **Configuration**: It provides interfaces to configure the console and the controls and allows the operator to check the system status (Figure 4.3).

4.3.3 Management

From *toolbar* by selecting *Management* button, it is possible to manage the city map according to the user profile.

Figure 4.3 CDSS GUI toolbar.

4.3.3.1 Map

User can view the city maps according to his profile. The *City Map* shows a 2D view of the monitored meter network, and each object is described with an appropriate icon.

On the left side of the screen, the user can access to some useful functionalities to work with the map:

- The tools to drag the map, to zoom in/out, and to calculate distances and areas;
- The legend of the icons used for the buildings on the map;
- The tool to filter the buildings by type and district (Figure 4.4).

Clicking on an icon showed on the map, the user obtains various detailed information on the chosen object (Figure 4.5).

User can click on the *Diagnostic* icon to check whether there are energy losses for that building, or on the *DataFlow Offline* icon. The *DataFlow Offline* informs that the building meter does not send data from more than 24 hours. In addition, the color of the icon suggests whether data are present or not:

- Green icon: No data are available.
- Red icon: Data are available (Figure 4.6).

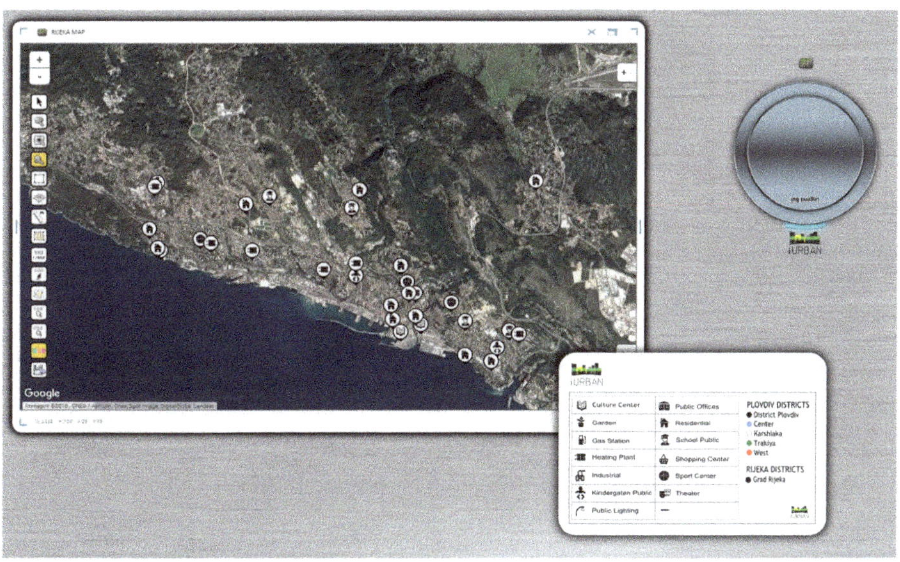

Figure 4.4 Management—map screenshot.

Figure 4.5 Management—map—building detailed information by clicking on the map on a specific icon.

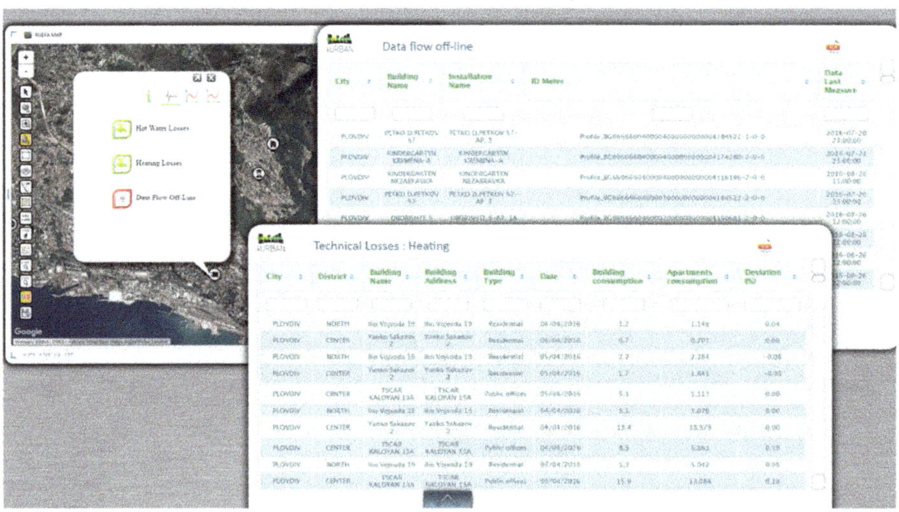

Figure 4.6 Management—map—diagnostic losses and DataFlow Offline.

User can click on the *Graph* icon and a short report on the consumption related to the last 24 hours, 7 days, and 30 days is available. It is possible to open the *Graph Filter Maker* window to check whether there are energy consumption and production data for that building, clicking on the *Chart Building* icon. This window is composed by three parts:

- *Filter Maker*
- *Graph Container*
- *Help Area*

The filters are already selected in order to show information related to the chosen object but the user can select filters according to his/hers profile:

- *Filter*
    - Energy type: electricity, heating, water, hot water, etc.
    - Level: city, district, building, installation (apartment), meters.
- *Time*
    - Daily, last week, last month, last year, and customizable range.
- *Chart*
    - EnergyView Consumption and EnergyView Production.

The *Graph Filter Maker* window provides the user a dashboard with different kinds of data aggregations relative to the energy consumption or production. The objective of this tool is to provide an overview of the energy management trend by several variables, as time, district, and energy type.

In particular, the dashboard is composed of the following charts:

- Monthly performance
- Average saving
- Quarters
- Week of month
- Total consumption (Figure 4.7).

4.3.4 CityEnergyView

From toolbar by selecting *CityEnergyView* button, it is possible to manage the *EnergyView* and *Consumption* 24H/7D/30D according to the user profile.

4.3.4.1 EnergyView

User can check the consumption and production energy data from *EnergyView*. User can select filter according to his/her profile. The system can show data at city, district, building, installation, and meter level and for different energy type (electricity, water, hot water, heating, etc.). Moreover, the user can select a period of time: daily, monthly, yearly, or a particular range of time.

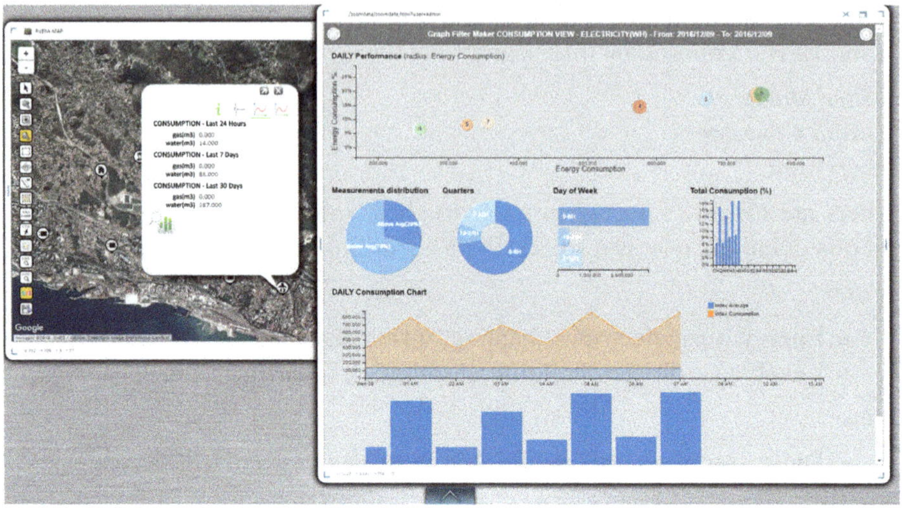

Figure 4.7 Graph filter maker window.

The *EnergyView* button opens a *Graph Filter Maker* window composed from three parts:

- Filter Maker
- Graph Container
- Help Area.

4.3.4.2 Filter Maker

This part allows the user to set the filters to use. The user can access it by clicking on the icon in the left upper corner. The *Filter Panel* is displayed and the user can choose the filters to apply between three buttons clicking on one of them:

- Filter
- Time
- Chart (Figure 4.8).

The *Filter* button allows the user to choose which cities, districts, buildings, installations, and/or meters and which energy type (electricity, water, heating, etc.) to select. After the filters choice, click *Apply* to create the charts (Figure 4.9).

The *Time* button allows the user to select temporal ranges. In this panel, the user can select from different default time ranges or a custom date. Once

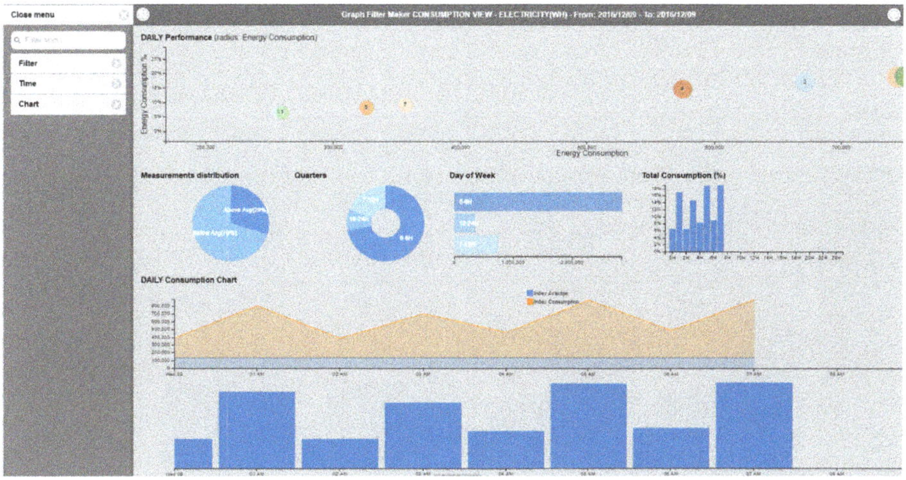

Figure 4.8 CityEnergyView—EnergyView—Filter Maker.

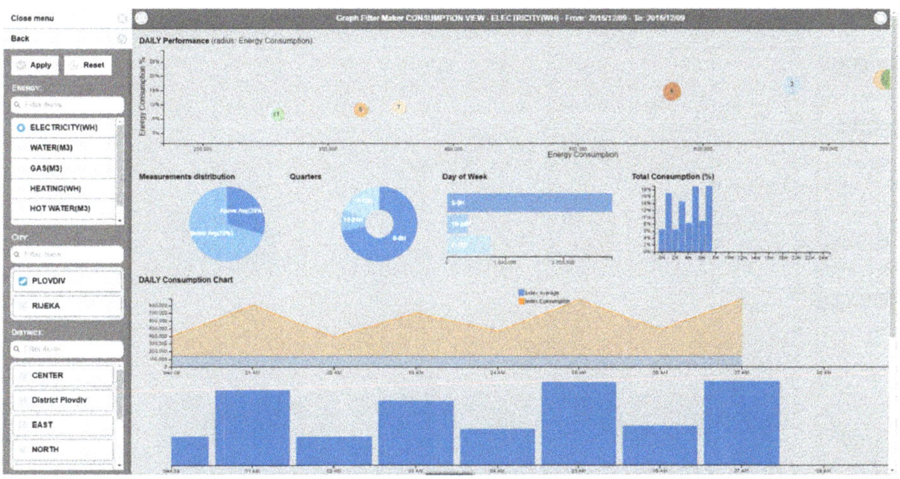

Figure 4.9 CityEnergyView—EnergyView—Filter button.

the time range is selected, the graphs will be showed into the *Graph Container* (Figure 4.10).

The *Chart* button allows the user to choose a chart view between EnergyView Consumption, EnergyView Production, and Energy Forecast (Figure 4.12).

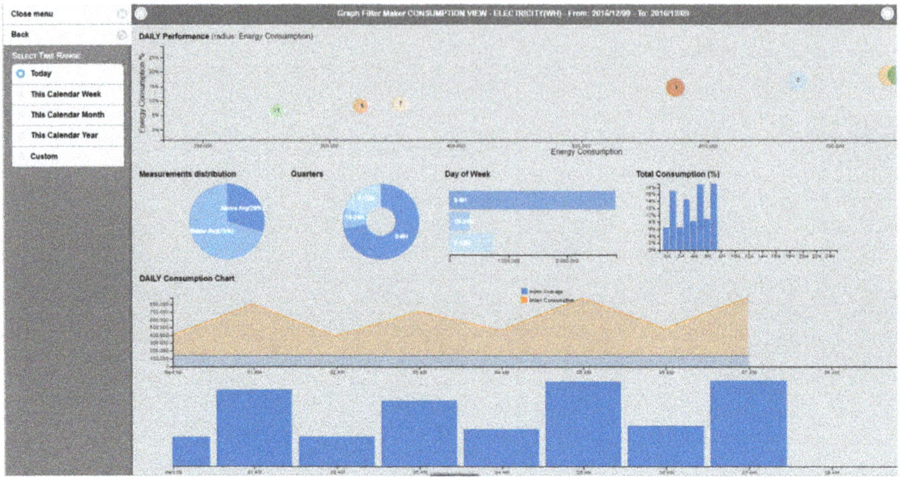

Figure 4.10 CityEnergyView—EnergyView—Filter Maker—Time button.

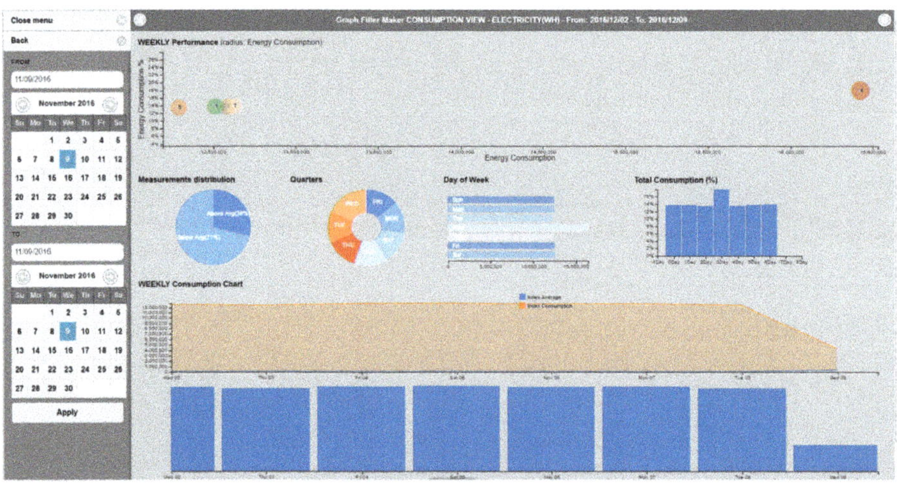

Figure 4.11 CityEnergyView—EnergyView—Filter Maker—Time button customizable range.

The user can visualize the graphs related to the consumption data by selecting the radio button *EnergyView Consumption* (Figure 4.13).

The user can visualize the graphs related to the production data by selecting the radio button *EnergyView Production* (Figure 4.14).

The user can visualize the forecast data by selecting the radio button *Energy Forecast* (Figure 4.15).

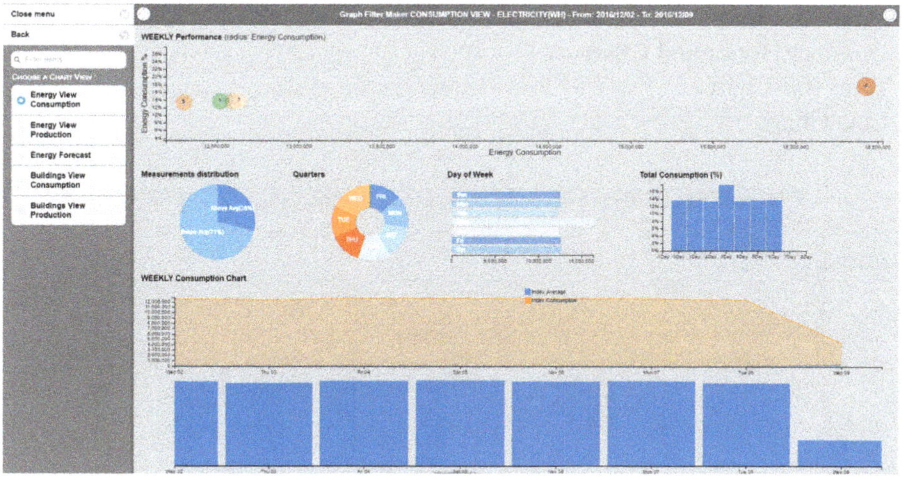

Figure 4.12 CityEnergyView—EnergyView—Chart button.

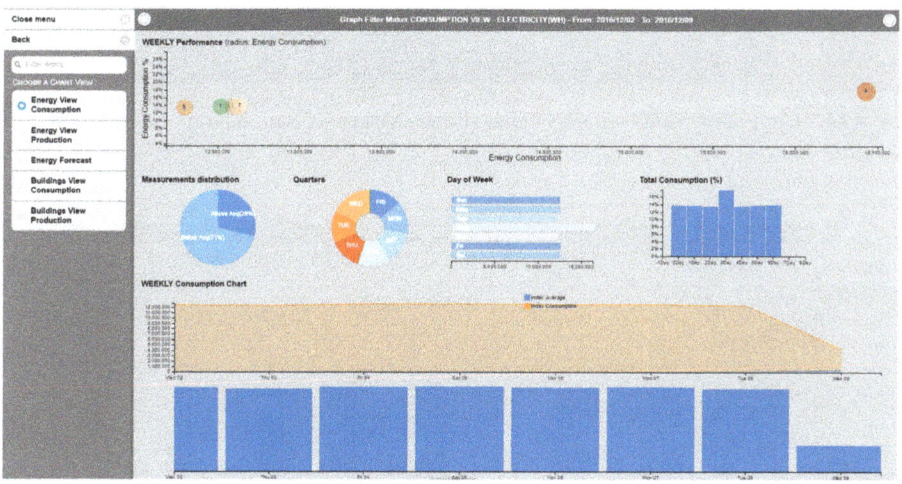

Figure 4.13 CityEnergyView—EnergyView—Filter Maker—Chart button: EnergyView consumption.

4.3.4.3 Graph Container

Graph Container is the hearth of the graphical component. It is in the middle of the page; here, six different charts, showing the filtered data, are presented:

- Bubble Chart
- Average Saving Chart

- Quarters Chart
- Quarters Horizontal Chart
- Total Bar Chart
- Line Chart.

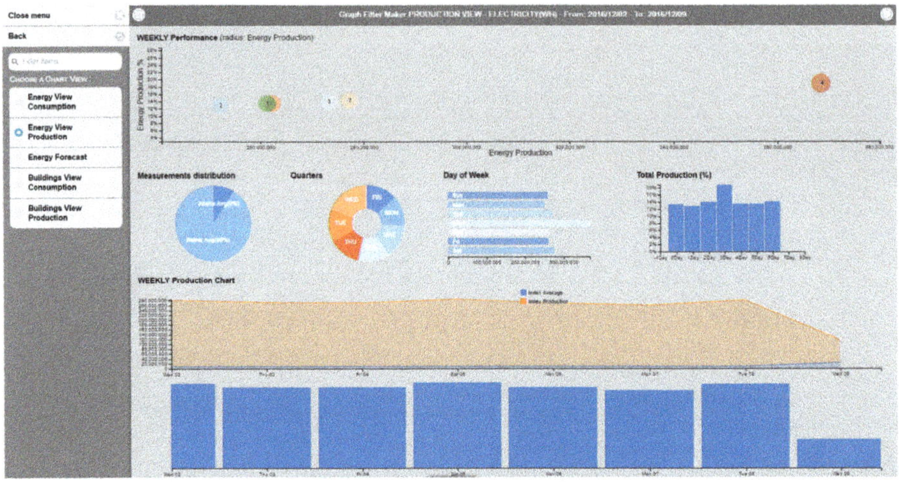

Figure 4.14 CityEnergyView—EnergyView—Filter Maker—Chart button: EnergyView production.

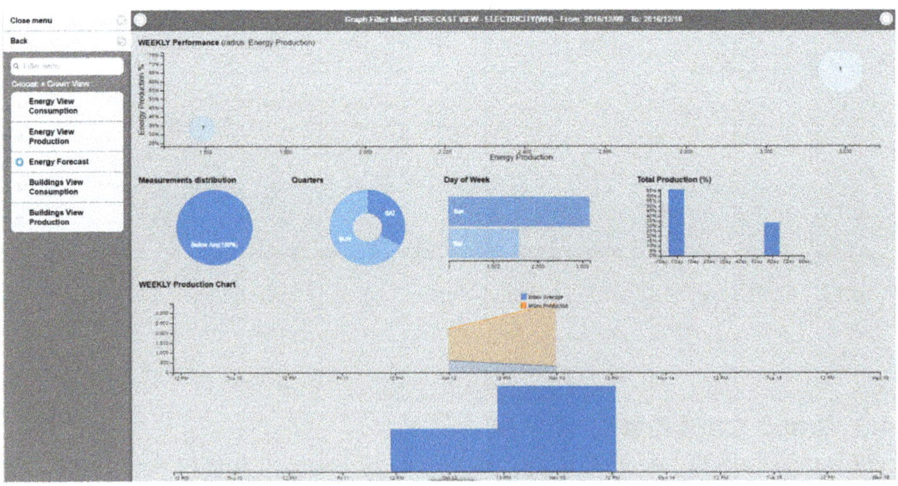

Figure 4.15 CityEnergyView—EnergyView—Filter Maker—Chart button: energy forecast.

The *Bubble Chart* is the first graph in the main container. It shows consumption and production as a percentage: the data are grouped in different time ranges. For example, by selecting "year" from the time filter, the graph view shows a representation of consumption or production grouped by month; instead, by selecting "month" from time filter, the graph view shows a representation of consumption or production grouped by week. Each radius bubble is the sum of values divided by the number of values in the list, and the result is multiplied by 100. Therefore, we can have the percentage of total consumption or production for each month. Months range is 1–12 (Figure 4.16).

The *Average Saving* Chart is a graph that shows energy savings or average of consumption and/or production in percentage.

For example, by selecting "Consumption" in to the "Chart filter," the saved percentage compared to the average value of consumption is presented. The consumption over the percentage provides consumption, and the consumption under the average gives a gap from average that representing savings (Figure 4.17).

The *Quarters* Chart is a graph that displays the production and consumption data of different time ranges in four parts (quarters, a quarter of month, a week, and a day). The value is expressed as a percentage.

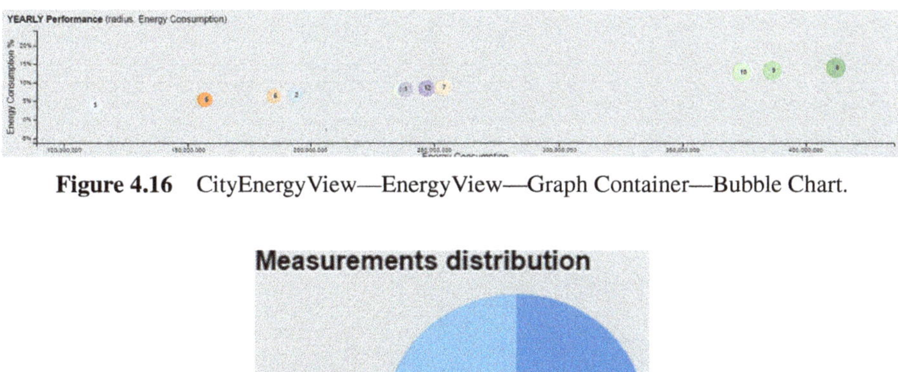

Figure 4.16 CityEnergyView—EnergyView—Graph Container—Bubble Chart.

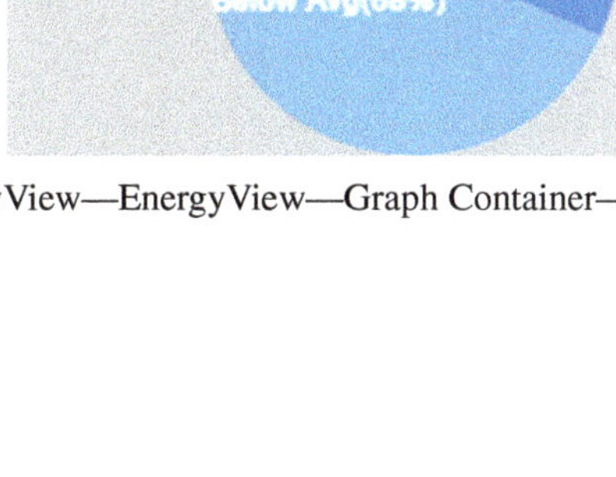

Figure 4.17 CityEnergyView—EnergyView—Graph Container—Measurements distribution.

For example, by selecting "year" from the time filter, the data are grouped by quarters. Each quarter is Q1 (January/February/March), Q2 (April/May/June), Q3 (July/August/September), and Q4 (October/November/December) (Figure 4.18).

The *Quarters Horizontal* Chart is a graph that is similar to the Pie Quarters Chart, but it shows the quarters in a horizontal way and represents values on the x-axis. For example, by selecting "year" from the time filter, the data are grouped by quarters. Each quarter is similar as above at the Quarters Chart (Figure 4.19).

The *Total Bar* Chart is a graph that displays the production and consumption data percentage in a stacked bar representation. For example, by selecting "year" from the time filter, the data are grouped by months. Each bar is a month identified by a number between 0 and 11 (Figure 4.20).

The *Line* Chart is at bottom of page, and it shows the production and consumption data line in the selected range.

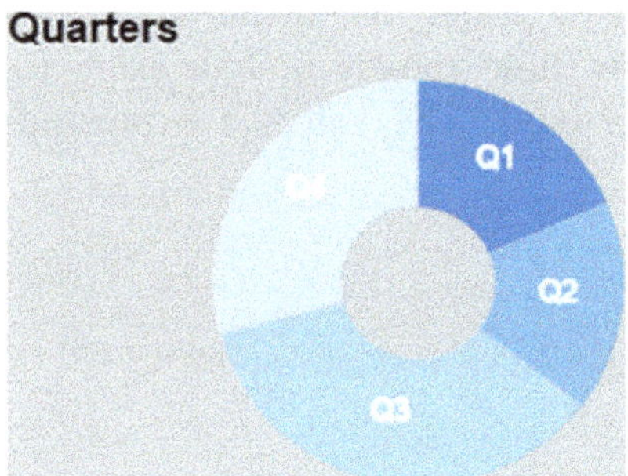

Figure 4.18 CityEnergyView—EnergyView—Graph Container—Quarters Chart.

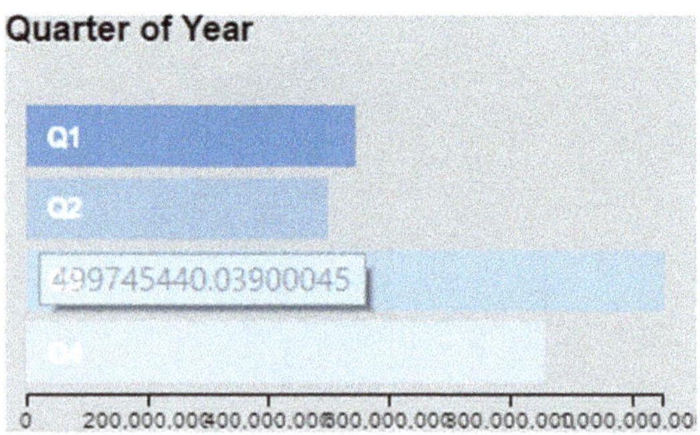

Figure 4.19 CityEnergyView—EnergyView—Graph Container—Quarters Horizontal Chart.

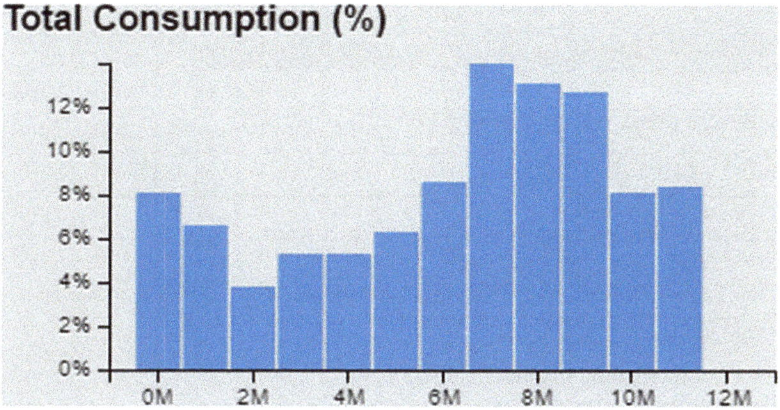

Figure 4.20 CityEnergyView—EnergyView—Graph Container—Total Bar Chart.

This graph has two lines: one for the average (blue) and one for the real data (orange). Beneath the x-axis, there is a bar chart that can elect a specific range. For example, by selecting "year" from the time filter, all values in a linear interpolation are showed (Figure 4.21).

4.3.4.4 Help Area

Help Area is on the upper right corner of display. This area helps the user with the *Help Pop-up* component. It provides a short description of each graph. This pop-up is ready to use when clicking on the icon in the upper right corner (Figure 4.22).

The Table 4.1 shows the relations between time filter and each grouping of graphs.

On the left column of the table, there are the "time" filters, and on the top row, there are the types of chart. For example, by selecting "Week" as time

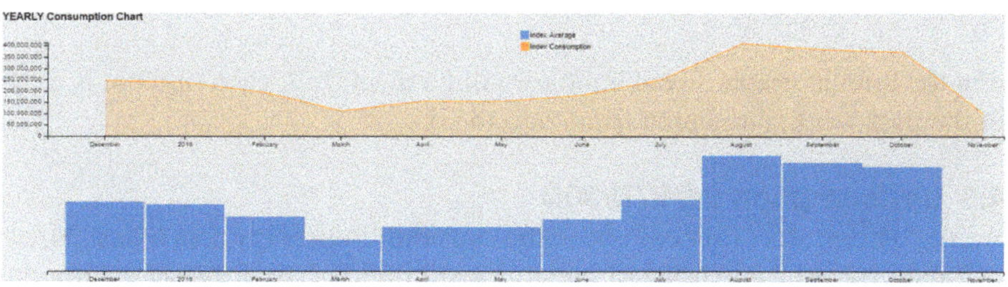

Figure 4.21 CityEnergyView—EnergyView—Graph Container—Line Chart.

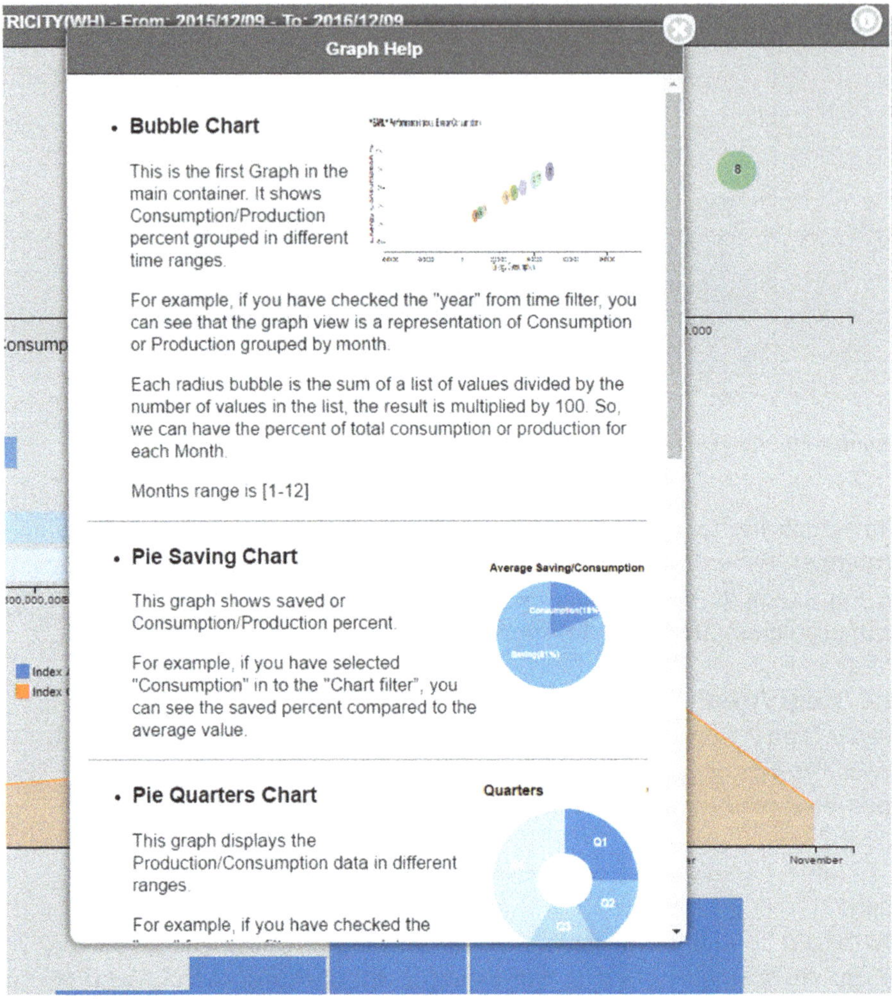

Figure 4.22 CityEnergyView—EnergyView—Graph Container—Help pop-up.

range in the bubble graph, seven bubbles will be displayed; each of them is a day of the past week, numbered from zero to six.

4.3.4.5 Consumption 24H/7D/30D

According to his profile, user can check the consumption energy data of last 24 hours, 7 days, or 30 days. A list of energy consumptions for different energy type (electricity, water, hot water, heating, etc.) at building level is showed.

Table 4.1

Grouped By	Bubble	Quarter	Horizontal Quarter	Total Bar	Line
Today	Hours of day	0–6 H = 0–6 hours 7–12 H = 7–12 hours 13–18 H = 13–18 hours 19–24 H = 19–24 hours	0–6 H = 0–6 hours 7–12 H = 7–12 hours 13–18 H = 13–18 hours 19–24 H = 19–24 hours	Hours of day	Hours of day
Week	Days of week	Days of week	Days of week	Days of week	Days of week
Month	Weeks of month	Weeks of month	Weeks of month	Days of month	Days of month
Year	Months of year	Q1 = 0–2 months Q2 = 2–5 months Q3 = 5–8 months Q4 = 8–11 months	Q1 = 0–2 months Q2 = 2–5 months Q3 = 5–8 months Q4 = 8–11 months	Months of Year	Months of Year

Consumption 24H

Consumption 24H opens a window that shows a list of the consumption energy data of last 24 hours. Consumption 24H list shows the following data:

- City: city name, related to the showed consumptions
- District: district name, related to the showed consumptions
- Building name: building name, related to the showed consumptions
- Energy type: electricity, water, heating, etc.
- Total consumption: total energy consumption, related to the energy type.

XLS icon opens and/or prints the report of the Consumption 24H list (Figure 4.23).

Consumption 7D

Consumption 7D opens a window that shows a list of the consumption energy data of last 7 days. Consumption 7D list shows the following data:

- City: city name, related to the showed consumptions
- District: district name, related to the showed consumptions
- Building name: building name, related to the showed consumptions
- Energy type: electricity, water, heating, etc.
- Total consumption: total energy consumption, related to the energy type.

XLS icon opens and/or prints the report of the Consumption 7D list (Figure 4.24).

iURBAN Total Consumption last 24 hours

City	District	Building Name	Energy Type	Total Consumption
PLOVDIV	CENTER	6 SEPTEMVRI 224	electricity(Wh)	390696.000
PLOVDIV	CENTER	Eliezer Kalev 12	electricity(Wh)	177784.200
PLOVDIV	CENTER	GEN.DANAIL NIKOLAEV 94	electricity(Wh)	153300.000
PLOVDIV	CENTER	HAN KUBRAT 8	electricity(Wh)	940082.768
PLOVDIV	CENTER	KINDERGARTEN NEZABRAVKA	electricity(Wh)	51398.000
PLOVDIV	CENTER	KINDERGARTEN PRIKAZEN SVIYAT	electricity(Wh)	139793.000
PLOVDIV	CENTER	KINDERGARTEN PRIKAZEN SVIYAT	gas(m3)	0.000
PLOVDIV	CENTER	PENCHO SLAVEJKOV 38	electricity(Wh)	475855.718
PLOVDIV	CENTER	PETKO D.PETKOV 57	electricity(Wh)	180481.000
PLOVDIV	CENTER	TSCAR KALOYAN 13A	electricity(Wh)	201042.000
PLOVDIV	CENTER	VESELA 33	electricity(Wh)	352956.000
PLOVDIV	CENTER	Yanko Sakazov 2	electricity(Wh)	949557.000
PLOVDIV	EAST	GEN. RADKO DIMITRIEV	electricity(Wh)	197100.000
PLOVDIV	EAST	GEN. RADKO DIMITRIEV 21	electricity(Wh)	2522494.000

Figure 4.23 CityEnergyView—Consumption 24H.

iURBAN Total Consumption last 7 days

City	District	Building Name	Energy Type	Total Consumption
PLOVDIV	CENTER	6 SEPTEMVRI 224	electricity(Wh)	3090780.000
PLOVDIV	CENTER	Eliezer Kalev 12	electricity(Wh)	1412532.000
PLOVDIV	CENTER	GEN.DANAIL NIKOLAEV 94	electricity(Wh)	1218000.000
PLOVDIV	CENTER	HAN KUBRAT 8	electricity(Wh)	7475589.709
PLOVDIV	CENTER	KINDERGARTEN NEZABRAVKA	electricity(Wh)	318522.000
PLOVDIV	CENTER	KINDERGARTEN PRIKAZEN SVIYAT	electricity(Wh)	778762.000
PLOVDIV	CENTER	KINDERGARTEN PRIKAZEN SVIYAT	gas(m3)	0.000
PLOVDIV	CENTER	PENCHO SLAVEJKOV 38	electricity(Wh)	3784030.715
PLOVDIV	CENTER	PETKO D.PETKOV 57	electricity(Wh)	1438066.000
PLOVDIV	CENTER	TSCAR KALOYAN 13A	electricity(Wh)	1598697.000
PLOVDIV	CENTER	VESELA 33	electricity(Wh)	2800332.000
PLOVDIV	CENTER	Yanko Sakazov 2	electricity(Wh)	7535892.000
PLOVDIV	EAST	GEN. RADKO DIMITRIEV	electricity(Wh)	1563300.000
PLOVDIV	EAST	GEN. RADKO DIMITRIEV 21	electricity(Wh)	19967350.000

Figure 4.24 CityEnergyView—Consumption 7D.

Consumption 30D

Consumption 30D opens a window that shows a list of the consumption energy data of last 30 days. Consumption 30D list shows the following data:

- City: city name, related to the showed consumptions
- District: district name, related to the showed consumptions
- Building name: building name, related to the showed consumptions
- Energy type: electricity, water, heating, etc.
- Total consumption: total energy consumption, related to the energy type.

XLS icon opens and/or prints the report of the Consumption 30D list (Figure 4.25).

4.3.5 Demand Response Management

From toolbar by selecting the *DR Management* button, it is possible to manage the DR program and the peaks monitoring according to the user profile. A DR program can be used to manage any consumption peaks related to any energy (two program kinds are managed: consumption and thermostat). It is composed

City	District	Building Name	Energy Type	Total Consumption
PLOVDIV	CENTER	6 SEPTEMVRI 224	electricity(Wh)	13298828.000
PLOVDIV	CENTER	Eliezer Kalev 12	electricity(Wh)	6061304.700
PLOVDIV	CENTER	GEN.DANAIL NIKOLAEV 94	electricity(Wh)	5221300.000
PLOVDIV	CENTER	HAN KUBRAT 8	electricity(Wh)	32029349.767
PLOVDIV	CENTER	KINDERGARTEN NEZABRAVKA	electricity(Wh)	1665192.000
PLOVDIV	CENTER	KINDERGARTEN PRIKAZEN SVIYAT	electricity(Wh)	3080806.000
PLOVDIV	CENTER	KINDERGARTEN PRIKAZEN SVIYAT	gas(m3)	0.000
PLOVDIV	CENTER	PENCHO SLAVEJKOV 38	electricity(Wh)	16221464.987
PLOVDIV	CENTER	PETKO D.PETKOV 57	electricity(Wh)	6169394.000
PLOVDIV	CENTER	TSCAR KALOYAN 13A	electricity(Wh)	6850116.000
PLOVDIV	CENTER	VESELA 33	electricity(Wh)	12008544.000
PLOVDIV	CENTER	Yanko Sakazov 2	electricity(Wh)	32276286.000
PLOVDIV	EAST	GEN. RADKO DIMITRIEV	electricity(Wh)	6708150.000
PLOVDIV	EAST	GEN. RADKO DIMITRIEV 21	electricity(Wh)	85707336.000

Figure 4.25 CityEnergyView—Consumption 30D.

of a number of actions to manage the demand of energy of the customer in response to the conditions proposed by the supplier, in order to reduce peak situations in the energy distribution grid. The system allows to manage a list of DR programs stored in the database and insert a new one. A DR program archive is designed in order to reuse a program for peaks of the same kind. From the DR program list (a detailed description is in the following "DR Program" paragraph), it is possible to create a DR program pressing the *New DR Program* button and filling the information required (a detailed description is in the following "DR Program" paragraph). A DR program is closely related to one or more actions that describe what the customer has to perform to fulfill the DR program that was sent. From the New DR program, it is possible to add one or more actions pressing the *Action* button and filling the information required (a detailed description is in the following "DR Program" paragraph).

Peaks monitoring manages the peak situations in the energy distribution grid identified by the system or manually inserted by the user. From peaks and DR monitoring (a detailed description is in the following "Peaks Monitoring" paragraph), it is possible to manually insert a peak pressing the *Add New Peak* button and filling the information required (a detailed description is in the following "Peaks Monitoring" paragraph). To reduce peak situations in the energy distribution grid, it is necessary to select a DR program to use, choosing among those previously entered, and link it to the peak from the peaks monitoring list (a detailed description is in the following "Peaks Monitoring" paragraph). When a peak was linked to a DR program, it is time to send the notifications to the customers. From peaks and DR monitoring, it is possible to send one or more notifications to the customers pressing the *Pencil* icon and filling the information required in the notification list (a detailed description is in the following "Peaks Monitoring" paragraph). Three buttons are proposed to the user in order to manage the notifications: create new notification, save new notification, and send notification to LDSS (a detailed description is in the following "Peaks Monitoring" paragraph). All notifications inserted that have not been sent by the user pressing the *Send Notification to LDSS* button will be sent anyway by the system at the scheduled time. When the action proposed to the customer is finished, the result related to the peak is presented in the notification list.

4.3.5.1 DR program

DR program opens a window that shows a list with the programs created from the system or from the user (Decision Maker, Supervisor). The DR program list shows the following data:

- DR program name: the name chosen for the DR program.
- Type: demand response type (thermostatdr or consumptiondr).
- Energy type: (electricity, water, heating, etc.).
- Tips: Suggestion should be followed by customer to achieve the goal required by DR program to improve the consumption or production.
- Reward: reward offered to entice the customer to change a behavior or habit.
- Creation date: creation date of the DR program.
- More icon opens a new pop-up that shows more information about DR program and related notifications.
- Delete icon shows whether a DR program can be deleted or not. The user can delete a DR program if there are not peaks or notification joined.

XLS icon opens and/or prints the report of the DR program list (Figure 4.26).

More icon shows the information related to the recommended user actions for a selected DR program. If no action is associated with a DR program, the Action Detail is empty.

In the upper rows of the pop-up is reported information related to the DR program, and in the lower rows, those related to the Action Detail.

The DR program presents the following fields:

- DR ID: demand response identifier.
- DR name: the name chosen for the DR program.

Figure 4.26 DR Management—DR Program.

- DR type: demand response type (thermostatdr or consumptiondr).
- Energy type: (electricity, water, heating, etc.).
- Creation date: creation date of the DR program.
- Tips: Suggestion should be followed by customer to achieve the goal required by DR program to improve the consumption or production.
- Reward: reward offered to entice the customer to change a behavior or habit.

The Action Detail presents the following fields:

- OfferID: It is the ID related to the offer.
- Action: It describes the behavior or the action to be taken by the user to get the reward.
- Action Operation: It indicates the value below or above the threshold that the user has to check. It is closely linked to Action Values.
- Action Value: Numeric value that indicates the threshold that the user has to check, in order to implement the right action. It is closely linked to Action Operation.
- Modify: Modify icon allows to change the action detail.
- Delete: Delete icon allows to delete the current action detail (Figure 4.27).

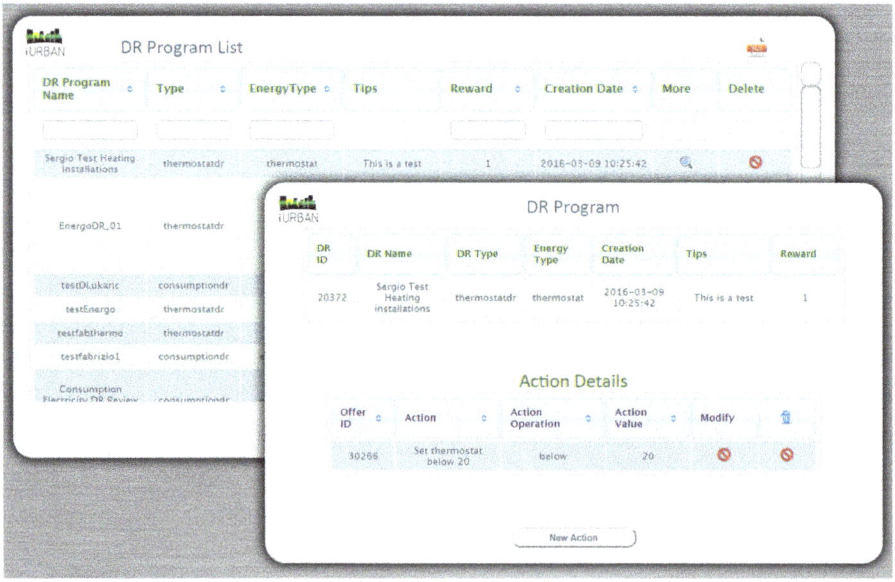

Figure 4.27 DR Management—DR Program—Action Detail.

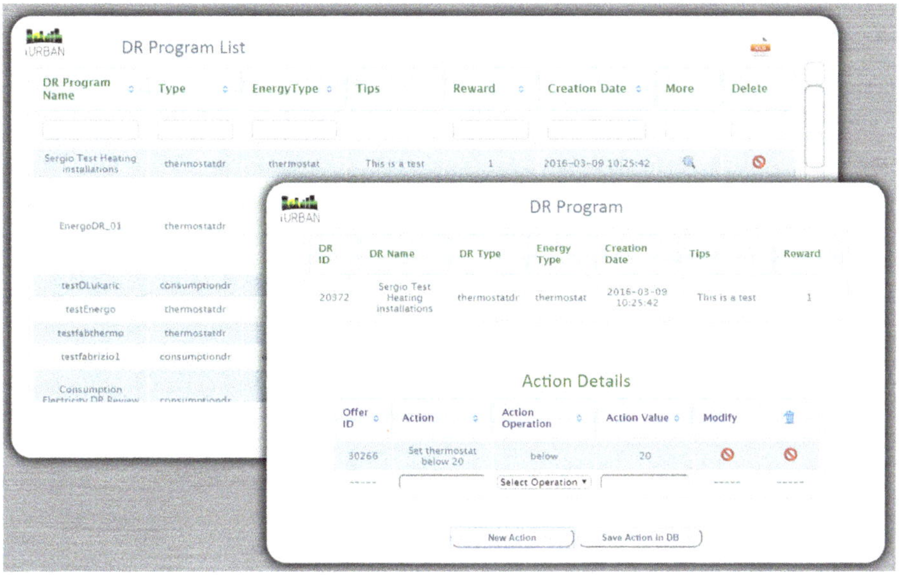

Figure 4.28 DR Management—DR Program—New Action.

New Action button presents a row under the current Action Detail in order to fill the fields to insert a new action.

Save Action in DB button saves in the CDSS DB the new action data inserted (Figure 4.28).

New DR Program button opens a pop-up where the user (Provider, Decision Maker, SuperUser) can insert a new DR program.

To insert, a DR program need the following fields:

- DR Program Name: mandatory
- DR Type: consumption, thermostat mandatory
- Energy Type: mandatory
- Reward: not mandatory
- Tips: not mandatory.

Action button opens a pop-up in order to insert economic offer related to the DR program created. Save button saves the new DR program (Figure 4.29).

With the *Action* button, the user (Provider, Decision Maker, SuperUser) can insert an action from this window filling the following fields:

- Action: action to be performed to get the benefit, mandatory.
- Action Operation: value action (below, above, etc.), mandatory.

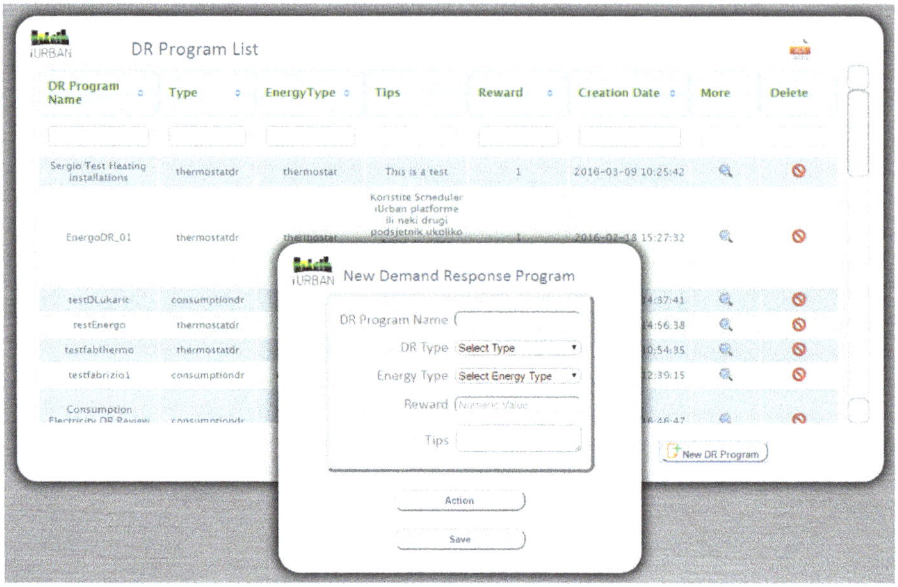

Figure 4.29 DR Management—DR Program—New DR Program button.

- Action Value: value of the action (numeric), mandatory.
- Delete: delete icon allows to delete the current action.

The *New Action* button adds a row to insert a new action. The *Save Action* button saves the new action inserted if the user saves the DR program (Figure 4.30).

4.3.5.2 Peaks monitoring

The peaks and DR monitoring allows the user to check the peaks generated if the energy production (response) is less than the energy consumption (demand). In this case, the user can try to resolve the problem using a suitable DR program. The peaks are calculated from the system or created manually from the user. It is possible to link a DR program with a peak. The fields showed in the "peaks and DR monitoring" list are the following:

- Type: user or system, inform the user whether the peak is manual or inserted by the system.
- City: city name where the peak should occur.
- Energy Type: kind of peak energy (electricity, water, heating, etc.).
- Start Time: start time of the peak.
- End Time: end time of the peak.

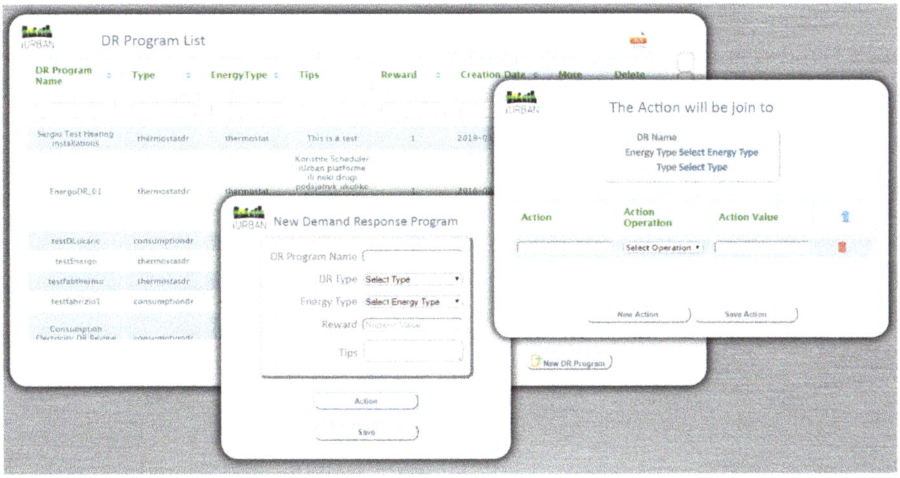

Figure 4.30 DR Management—DR Program—New DR Program—New Action.

- DR Program: It is the DR program linked to the peak. The field presents a list of DR program stored in the database, by the system or by the user, and related to the energy type of the peak.
- Status Icon: current peak status. It opens a pop-up with more information about peak status.
- Notification Icon: It opens a window in order to create and manage the notifications to the customer for the DR program selected (Figure 4.31).

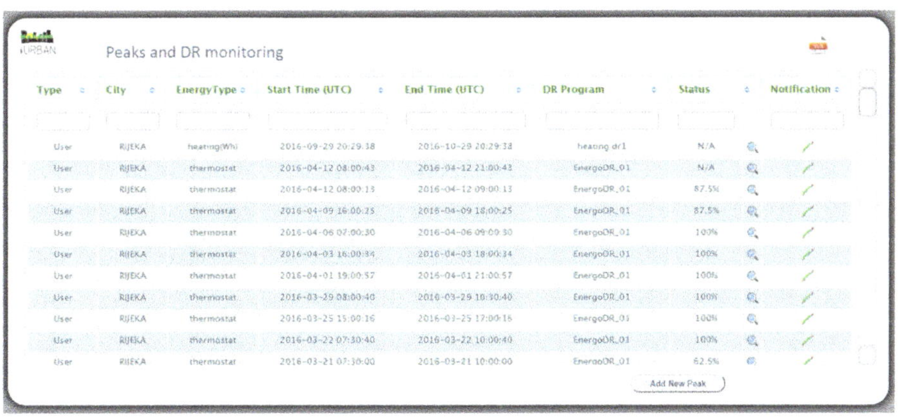

Figure 4.31 DR Management—Peaks and DR Monitoring.

Add New Peak button allows the user to insert a new peak (manual action) adding a row below the peak list (Figure 4.32).

The fields showed in the "Status Detail" list are the following:

- Group: associated group.
- Start Time: start time of the peak.
- Expiration Time: expiration time of the peak.
- Action Value: value of the action (numeric).
- Action Operation: value action (below, above, etc.).
- Installation ID: installation identifier.
- Installation Name: name of the installation.
- Results: results obtained (Figure 4.33).

The fields showed in the "Notification List" are the following:

- Group: customers for which the notification is applied.
- Scheduled Time: scheduled time of the peak.
- Start Time: start time of the peak.
- Expiration Time: expiration time of the peak.
- Action: action to be performed to get the benefit.
- Action Value: value of the action (numeric).
- Action Operation: value action (below, above, etc.).
- Notification Status: status of the notification.
- Results: results obtained.

Create new notification button allows the user to insert a new notification adding a row below the notification list. *Save new notification* button allows

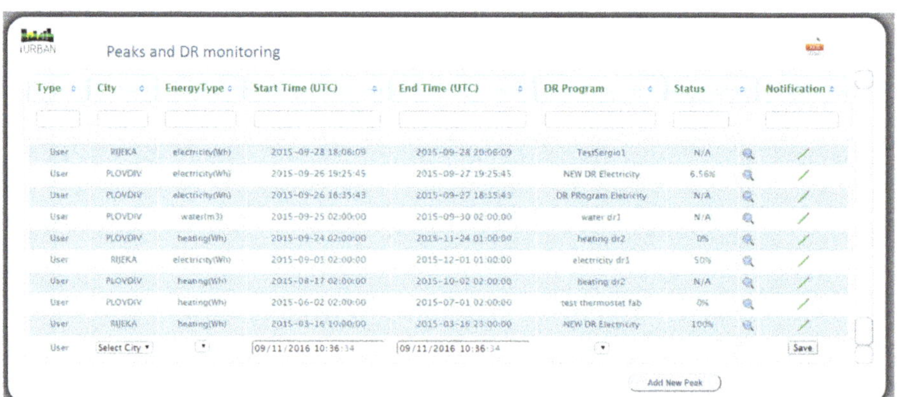

Figure 4.32 DR Management—Peak and DR Monitoring—Add New Peak.

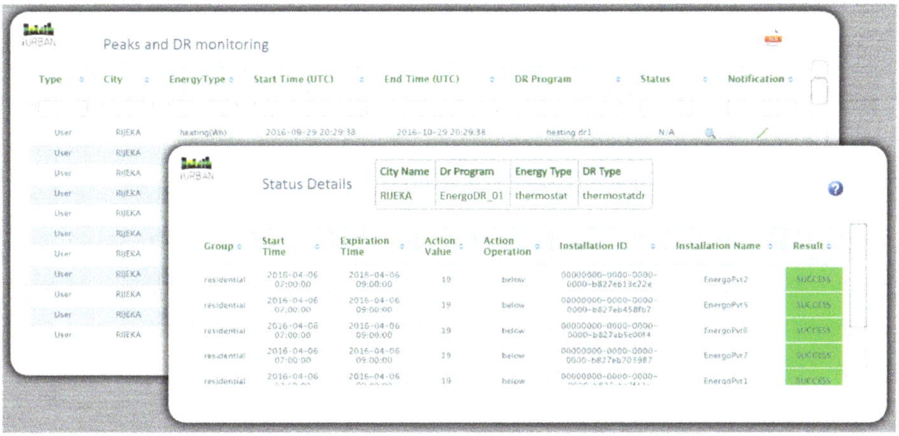

Figure 4.33 DR Management—Peaks and DR Monitoring—Status Detail.

the user to save the new notification. *Send notification to LDSS* button allows the user to send to LDSS one or more notifications previously selected in the checkbox on the left side of the list. If the user does not send the notification manually, it will be sent anyway at scheduled time by the system (Figure 4.34).

4.3.6 Tariff

From toolbar by selecting tariff button, it is possible to manage the tariff plans and the tariff comparison according to the user profile.

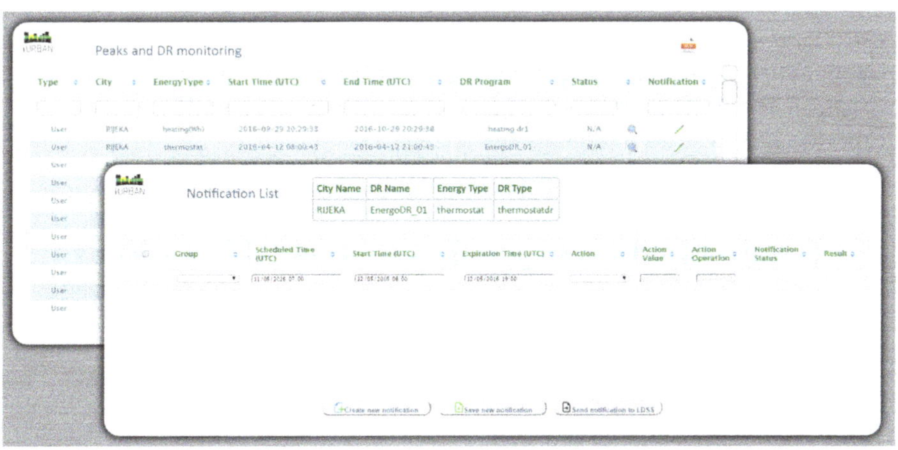

Figure 4.34 DR Management—Peaks and DR Monitoring—Notification List.

Figure 4.35 Tariff—Tariff Plans.

4.3.6.1 Tariff Plans

Tariffs Plans manage the tariffs applied to the customers. It is possible to insert new tariff and manage data related to the dynamic tariffs.

Tariffs Plans list presents the following information:

- City: city name where tariff is applied.
- Tariff Name: the name of the tariff.
- Energy Type: energy type related to the tariffs plans.
- Energy Unit: energy measure unit.
- Currency: currency applied.
- Group: customers for which the tariff is applied.
- Creation Date: start range for which the tariff plan is applied.
- Edit: the icon allows to change the tariff data.
- Delete: icon to delete the tariff.

XLS icon opens and/or prints the report of the tariff plan list.

The user can visualize the data applying a filter on each field (Figure 4.35).

New Tariff Plan button opens a pop-up where the user (Provider, Decision Maker, SuperUser) can insert a new tariff plan.

The mandatory fields are the following:

- City: the city name where apply the tariff.
- Tariff Name: the name of the tariff plan.
- Group: customer group to whom the tariff plan should be applied.

- Commodity: raw material.
- Unit: measure unit.
- Currency: the currency applied.

For each tariff, the user can create a tariff plan composed of one or more level (for the dynamic tariff for each level, it is possible to select a time of validity).

The fields to insert for each level are the following:

- ID level: level identifier.
- Level name: name of the level.
- Start: validity start date of the tariff plan (dd/mm/yy).
- End: validity end date of the tariff plan (dd/mm/yy).
- allDay: as an alternative to the start date and end date.
- Currency: the currency applied.
- Cost: cost for unit measurement.
- Edit icon: allows to change the tariff data.
- Delete icon: deletes the row.

The user can add new level click-on *Add Level* button that adds new line in the list. To save the tariff plan, the user can click-on *Save Tariff* button. XLS icon opens and/or prints the report of the tariff plan. On the base of the tariff applied, it will be possible to calculate the bill for each customer (Figure 4.36).

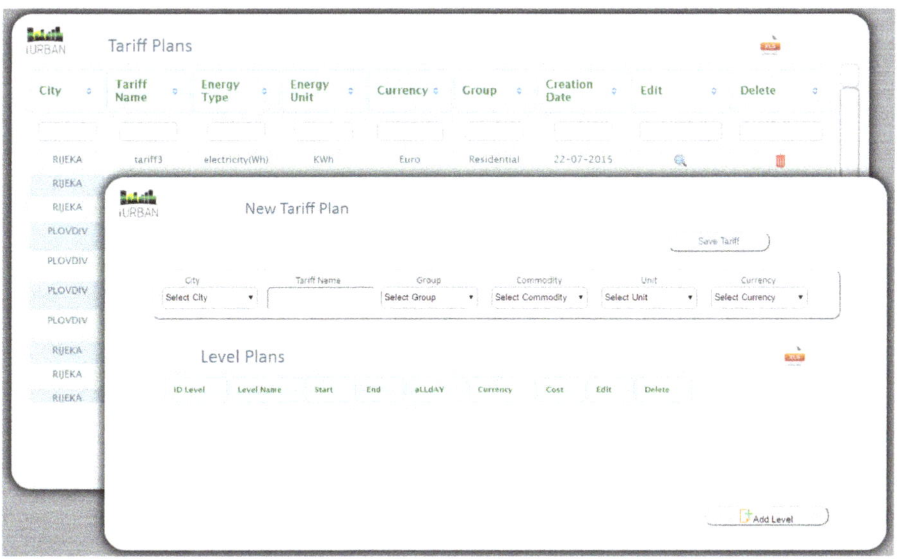

Figure 4.36 Tariff—Tariff Plans—New Tariff Button.

4.3.6.2 Tariff comparison

The tariff comparison shows a bar chart representing the comparison between some tariffs applied to the consumptions of the last year (Figure 4.37).

In the bar chart, the x-axis represents the months of the year and the y-axis represents the costs applied.

4.3.7 Diagnostic

From toolbar by selecting the *Diagnostic* button, it is possible to manage the DataFlow Offline and the technical losses according to the user profile.

4.3.7.1 DataFlow Offline

DataFlow Offline shows the list of meters (ID Meters) which do not send information to the system from more than 24 hours. That information can be useful in order to verify the status of the meters.

The DataFlow Offline shows the data for the cities and energy type related to the user profile.

Figure 4.37 Tariff—Tariff Comparison—Consumption/Price Curve.

Figure 4.38 Diagnostic—DataFlow Offline.

The DataFlow Offline table gives the following information:

- City: city name.
- Building Name: address of the building or indication of building type (e.g., school, CHP Plant).
- Installation Name: address of the building or indication of building type and floor or apartment number of the installation.
- ID Meter: Id alphanumeric of the meter.
- Data Last Measure: data of the last measure stored on the database.

The user can visualize the data applying a filter on each field.

XLS icon opens and/or prints the report of the DataFlow Offline list (Figure 4.38).

4.3.7.2 Hot Water Technical Losses

Hot Water Technical Losses is a loss of hot water in apartment buildings. It is the deviation calculated comparing the hot water provided to the entire building with hot water consumptions of each apartment of the building. The data of losses presented are described in the following:

- City: name of the city where the building is situated.
- District: name of the district where the building is situated.

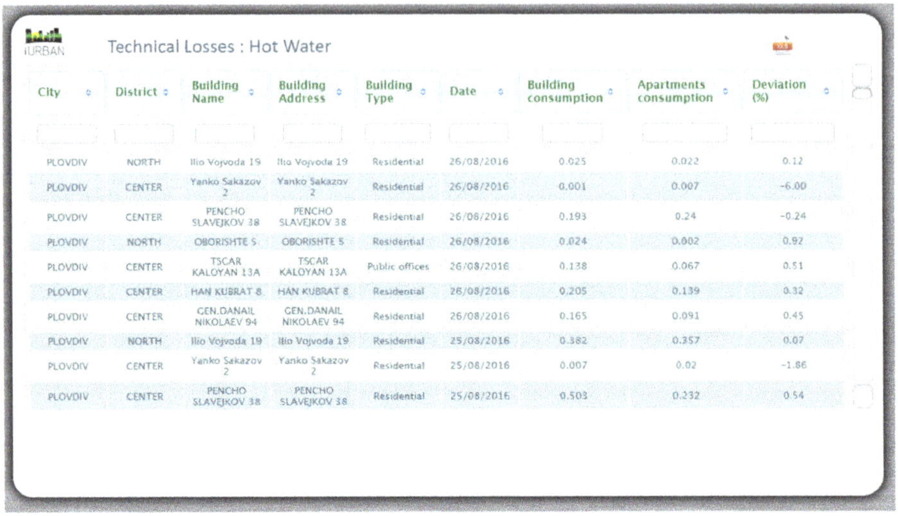

Technical Losses : Hot Water

City	District	Building Name	Building Address	Building Type	Date	Building consumption	Apartments consumption	Deviation (%)
PLOVDIV	NORTH	Ilio Vojvoda 19	Ilio Vojvoda 19	Residential	26/08/2016	0.025	0.022	0.12
PLOVDIV	CENTER	Yanko Sakazov 2	Yanko Sakazov 2	Residential	26/08/2016	0.001	0.007	-6.00
PLOVDIV	CENTER	PENCHO SLAVEJKOV 38	PENCHO SLAVEJKOV 38	Residential	26/08/2016	0.193	0.24	-0.24
PLOVDIV	NORTH	OBORISHTE 5	OBORISHTE 5	Residential	26/08/2016	0.024	0.002	0.92
PLOVDIV	CENTER	TSCAR KALOYAN 13A	TSCAR KALOYAN 13A	Public offices	26/08/2016	0.138	0.067	0.51
PLOVDIV	CENTER	HAN KUBRAT 8	HAN KUBRAT 8	Residential	26/08/2016	0.205	0.139	0.32
PLOVDIV	CENTER	GEN.DANAIL NIKOLAEV 94	GEN.DANAIL NIKOLAEV 94	Residential	26/08/2016	0.165	0.091	0.45
PLOVDIV	NORTH	Ilio Vojvoda 19	Ilio Vojvoda 19	Residential	25/08/2016	0.382	0.357	0.07
PLOVDIV	CENTER	Yanko Sakazov 2	Yanko Sakazov 2	Residential	25/08/2016	0.007	0.02	-1.86
PLOVDIV	CENTER	PENCHO SLAVEJKOV 38	PENCHO SLAVEJKOV 38	Residential	25/08/2016	0.503	0.232	0.54

Figure 4.39 Diagnostic—Technical Losses: Hot Water.

- Building name: name of the building where the loss occurs.
- Building address: address of the building where the loss occurs.
- Building type: kind of the building (sport center, public building, residential building, etc.).
- Date: date of the loss detection.
- Building consumption: total building consumption.
- Apartments consumption: total apartments consumption.
- Deviation: Deviation in percentage calculated by the system (Figure 4.39).

4.3.7.3 Heating Technical Losses

Heating Technical Losses is a loss of heating in apartment buildings. It is the deviation calculated comparing the heating provided to the entire building with heating consumptions of each apartment of the building.

The data of losses presented are described in the following:

- City: name of the city where the building is situated.
- District: name of the district where the building is situated.
- Building name: name of the building where the loss occurs.
- Building address: address of the building where the loss occurs.

- Building type: kind of the building (sport center, public building, residential building, etc.).
- Date: date of the loss detection.
- Building consumption: total building consumption.
- Apartments consumption: total apartments consumption.
- Deviation: deviation in percentage calculated by the system (Figure 4.40).

4.3.8 Weather Forecast

From the toolbar by selecting *Weather Forecast* button, it is possible to see the weather information related to the city of interest (Figure 4.41).

4.3.9 User

From the toolbar by selecting *User* button, it is possible to manage the user login and logout according to the user profile. A pop-up is shown containing all the information on the current user; it is possible to logout with the user, clicking on the icon on the upper right corner (Figure 4.42).

iURBAN Technical Losses : Heating

City	District	Building Name	Building Address	Building Type	Date	Building consumption	Apartments consumption	Deviation (%)
PLOVDIV	NORTH	Ilio Vojvoda 19	Ilio Vojvoda 19	Residential	06/04/2016	1.2	1.149	0.04
PLOVDIV	CENTER	Yanko Sakazov 2	Yanko Sakazov 2	Residential	06/04/2016	0.7	0.703	0.00
PLOVDIV	NORTH	Ilio Vojvoda 19	Ilio Vojvoda 19	Residential	05/04/2016	2.2	2.384	-0.08
PLOVDIV	CENTER	Yanko Sakazov 2	Yanko Sakazov 2	Residential	05/04/2016	1.7	1.841	-0.08
PLOVDIV	CENTER	TSCAR KALOYAN 13A	TSCAR KALOYAN 13A	Public offices	05/04/2016	5.1	5.113	0.00
PLOVDIV	NORTH	Ilio Vojvoda 19	Ilio Vojvoda 19	Residential	04/04/2016	5.1	5.076	0.00
PLOVDIV	CENTER	Yanko Sakazov 2	Yanko Sakazov 2	Residential	04/04/2016	13.4	13.379	0.00
PLOVDIV	CENTER	TSCAR KALOYAN 13A	TSCAR KALOYAN 13A	Public offices	04/04/2016	8.5	6.863	0.19
PLOVDIV	NORTH	Ilio Vojvoda 19	Ilio Vojvoda 19	Residential	03/04/2016	5.3	5.042	0.05
PLOVDIV	CENTER	TSCAR KALOYAN 13A	TSCAR KALOYAN 13A	Public offices	03/04/2016	15.9	13.084	0.18

Figure 4.40 Diagnostic—Technical Losses: Heating.

Figure 4.41 Weather Forecast.

Figure 4.42 User—User Information Pop-up.

4.3.10 Configuration

The *Configuration* button of the toolbar manages the console and the controls according to the user profile.

4.3.10.1 Console

The *Console* window manages the iURBAN console adding, modifying, or deleting a button on the toolbar (Figure 4.43).

Clicking on the *button name*, listed on the left of the window, all the information related to it will be showed on the right. It is possible to modify or delete it (Figure 4.44).

Selecting *New Button*, the fields to be filled to insert a new button will be presented on the right of the window (Figure 4.45).

4.3.10.2 Controls

From the *Controls* window, it is possible to add, modify, or delete the iURBAN Controls (Figure 4.46).

Clicking on the control name, listed on the left of the window, all the information related to it will be showed on the right (Figure 4.47).

Figure 4.43 Configuration—Console.

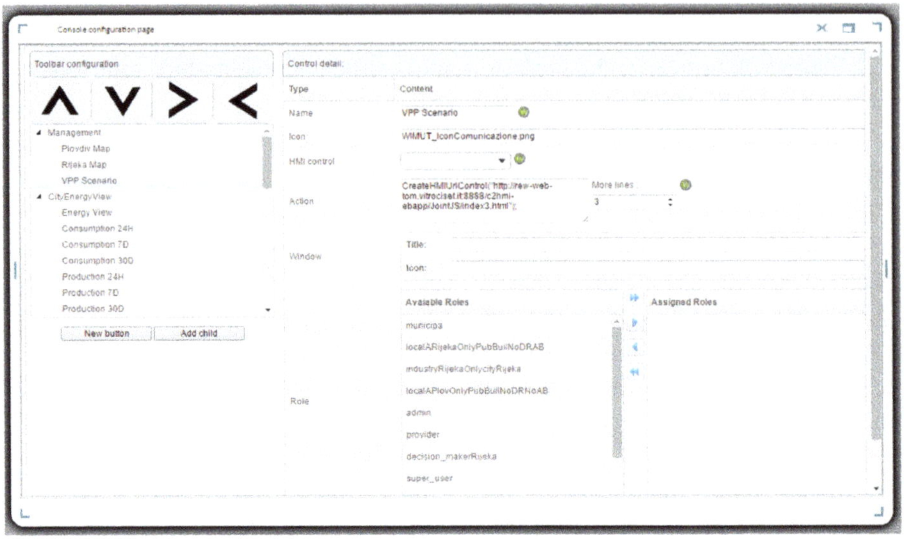

Figure 4.44 Configuration—Console—Button Configuration.

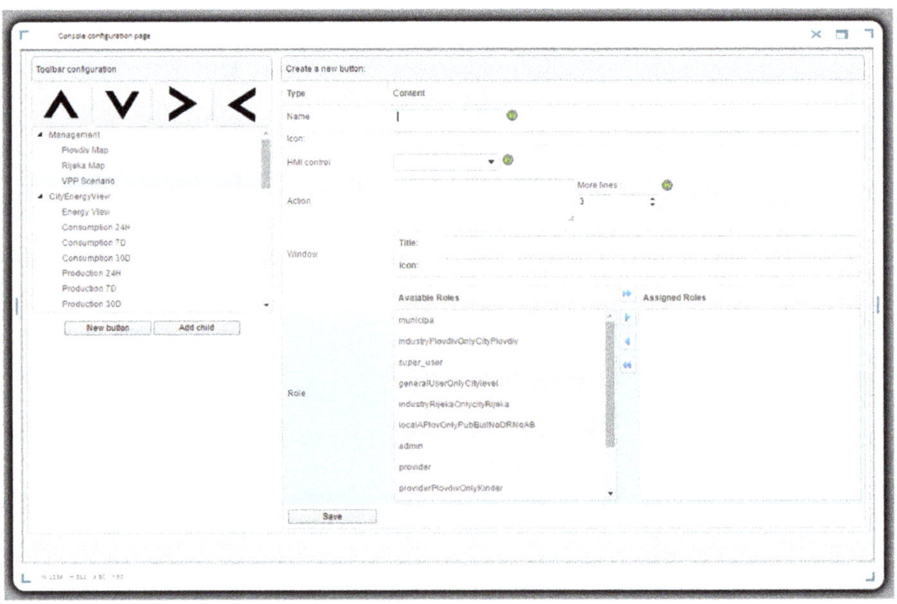

Figure 4.45 Configuration—Console—New Button.

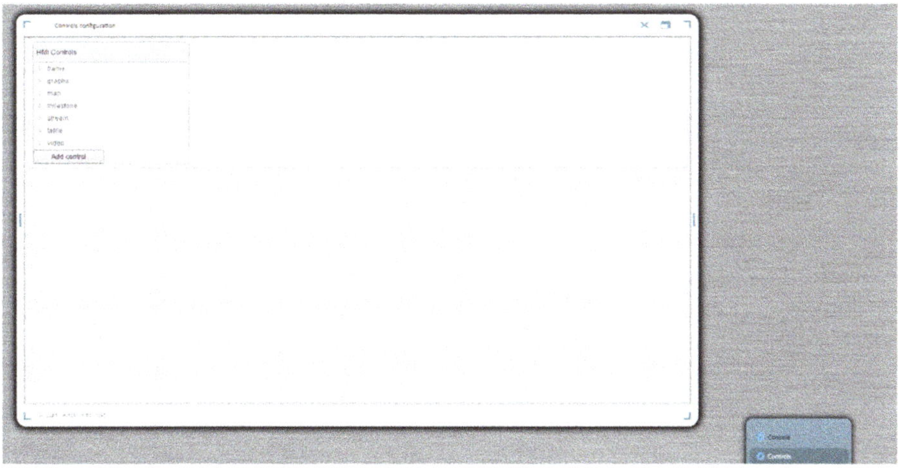

Figure 4.46 Configuration—Controls.

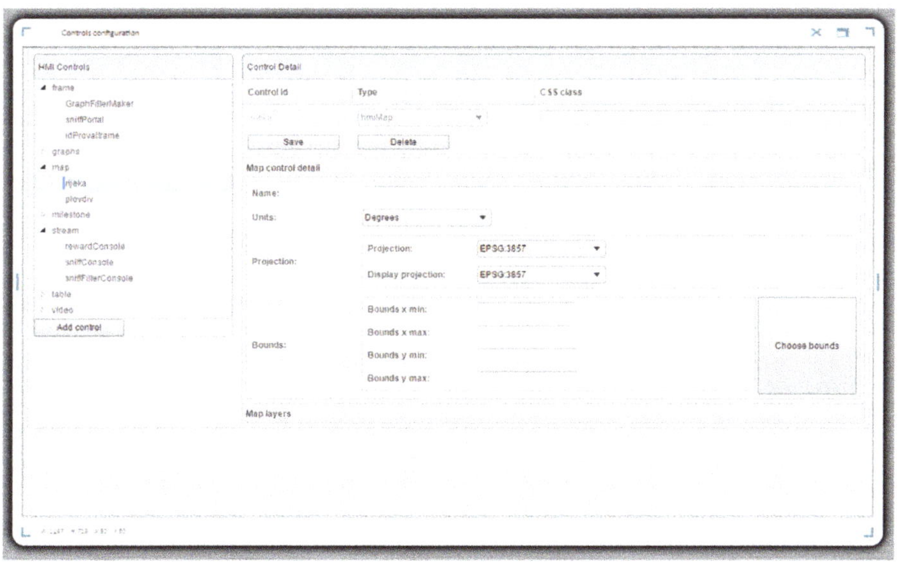

Figure 4.47 Configuration—Controls—Control Data.

4.4 Conclusion

The CDSS aggregates data from all LDSSs to provide city-level decision support to authorities and energy service providers. The CDSS generates a number of parameters, including citywide energy production and consumption forecasts. CDSS allows the users to do the following:

- Get a continuous snapshot of city energy consumption and production;
- Manage energy consumption and production;
- Forecast energy consumption;
- Plan new energy "producers" for the future needs of the city;
- Visualize, analyze, and take decisions of all the end points that are consuming or producing energy in a city level, permitting them to forecast and planning renewable power generation available in the city, a real-time optimization and being perfectly scalable (meaning its ability to be enlarged to accommodate that growth of data).

Analyzing the results gathered in the two pilots of iURBAN project, we can say that the user interface and the interaction paradigm implemented for CDSS are really innovative compared with the systems that usually people from energy utilities and municipalities are used to work with. However, the functionalities and the data reports provided by CDSS must be improved in order to cover all the needs of the users from energy utilities and municipalities. The architecture of CDSS will allow us to easily extend it with new features following the new requirements indicated from the users.

5

iURBAN LDSS

Alberto Fernandez

Sensing & Control, Barcelona, Spain

Abstract

This chapter provides an overview of the iURBAN graphical user interface developed targeting households. Provides a summary of functions supported and its corresponding interface.

Keywords: Smart home, Graphical user interface, Energy management, Energy visualization, Demand response, Notifications, Messaging, iOS, Android.

5.1 Introduction

The local decision support system component of iURBAN, called LDSS, is responsible to deliver a set of functions to households. It is composed by a back end and front end based on Web and mobile phone-based GUI. Figure 5.1 shows the component in relation to the Smart City Database (SCDB) and the centralized decision support system (CDSS).

LDSS main goal is to engage consumers and prosumers on the efficient use of energy. The engagement is based on data, captured in near real-time, related to their energy consumption, as well as energy production from their installed distributed energy resources (DER). The engagement is target throughout a user-friendly interface using every-day-use devices such as smartphones, tablets, and PCs. The LDSS is the main tool connecting households with their energy footprint and energy services. LDSS is built on the goal to offer a complete set of smart home functionalities grouped in four main blocks:

- Comfort
    - Climate control

    - Comfort levels (temperature, humidity, and luminance)
    - Air quality
- Energy management
- Security
    - Smoke, carbon oxide (CO), flood detection
    - Movement detectors
    - Magnetic detectors (for doors, windows)
    - Arm and disarm function
- Automation
    - Lights
    - Smart plugs
    - Thermostats

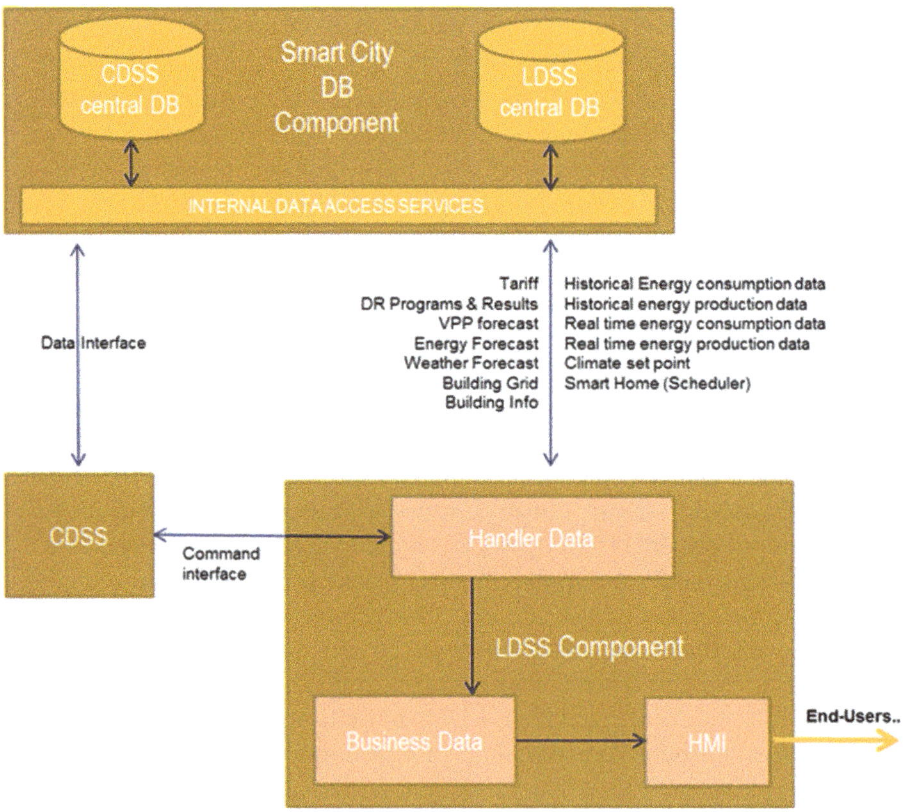

Figure 5.1 LDSS component and its relations.

iURBAN LDSS user installations provide data through a set of sensors and actuators based on Z-wave and wireless MBUS pulse reader off-the-shelve products and/or smart meters. The interface can be configured so the user can jump directly to energy information in case they do not have installed other devices than the smart meter (i.e., they do not have smart home enabled functionalities).With regard to energy management, the following features are offered:

- Visualization and exploration of energy consumed and produced
    - Historical data
    - Real-time data
    - Comparison
    - Export
- Visualization of energy consumption and production prediction
- Energy visualized
    - Electricity, gas, water, and heating
- Categorization of electricity consumption by means of retrofitting
- User engagement
    - Tips
    - Energy consumption/production comparison with neighborhood.
    - Personalized messages
    - Rewards
- Weather forecast
- Demand response
    - Heating
    - Electricity

5.2 Graphical User Interface

As the main tool to engage households on energy efficiency, LDSS is responsible for combining energy consumption and production visualization with user engagement.

The graphical design took into account different user levels of interest in relation with energy consumption/production and energy efficiency. At outer level, the tool visualizes information using a simple red/yellow/green color scheme. Red color is used when over consumption or inefficient consumption is detected by the artificial intelligence modules of iURBAN; on the opposite, green color is used to inform users' good efficiency on energy consumption if achieved at their home.

From the outer level, the user can dig down on to obtain detailed information about the consumptions/production and forecast, for instance, is able to plot energy consumption/production per hour/day/week/month basis, or even raw data captured by smart meters. The rich set of features associated with energy exploration makes the LDSS a powerful tool to explore and understand the reasons of the energy consumed/produced at home.

LDSS has been also developed taking into account needs of service providers (utilities, energy retailers, telecommunication companies, security companies, facility managers, etc.). LDSS brings a powerful mechanism for them to communicate with clients. Initially target to issue tips to improve energy efficiency, however the mechanism can be extended to send personalized messages to each client. This communication mechanism helps service provider to keep customer loyalty.

Enabling the development of new energy services is easy through the infrastructure provided by iURBAN and the LDSS. LDSS provides an example of new energy services as a way to prove it. The service that has been implemented is demand response programs for climate control and electricity consumption.

LDSS, through different means of communication channels such as SMS, e-mail, and push notifications to iOS and Android devices, allows service provider to notify demand response actions to their customers. Upon reception, LDSS user can modify thermostat settings and energy consumption to achieve the requests. Current LDSS version does not support automatic settings of smart plugs nor climate controllers, due to the context of execution of the pilots; however, automatization at this level can be achieved with extremely low effort. Specific user interfaces have been developing providing users about the benefits of achievement of the demand response as well as how to achieve it.

LDSS has been designed targeting Web and smartphone access. In case of Web interface, LDSS has been designed using the following:

- ASP.NET MVC.[1] The Web server has been deployed in Azure Web Sites.[2]

In case of smartphone, LDSS has been designed for the following:

- Android API version 16 or above.
- iOS 7.0 or above.

---

[1] http://www.asp.net/mvc.

[2] https://msdn.microsoft.com/en-us/magazine/jj883953.aspx.

5.2.1 Main Graphical User Interface Functionalities

Within this section, the LDSS graphical user interface is explained. The next figures are based on the Web interface, but the reader should take into account that the same principles apply for iOS and Android apps, so the user is finding the same functions and visualizations in the entire ecosystem of visualization.

The service is accessible at https://ServiceURL/Account/Login as shown in Figure 5.2.

The graphical user interface has been designed to be very simple and friendly. Special effort has been done to represent energy consumption (and production when available) to LDSS users.

Energy information is shown divided into three periods of time by default: (i) day, (ii) week, and (iii) month as shown in Figure 5.3(b). If user is interested in other means of energy exploration, the interface allows them to select different periods of time.

Energy information is shown for the four types of energy being provided by iURBAN utilities and energy agencies: (i) electricity, (ii) heating, (iii) gas, and (iv) water as shown in Figure 5.3(a).

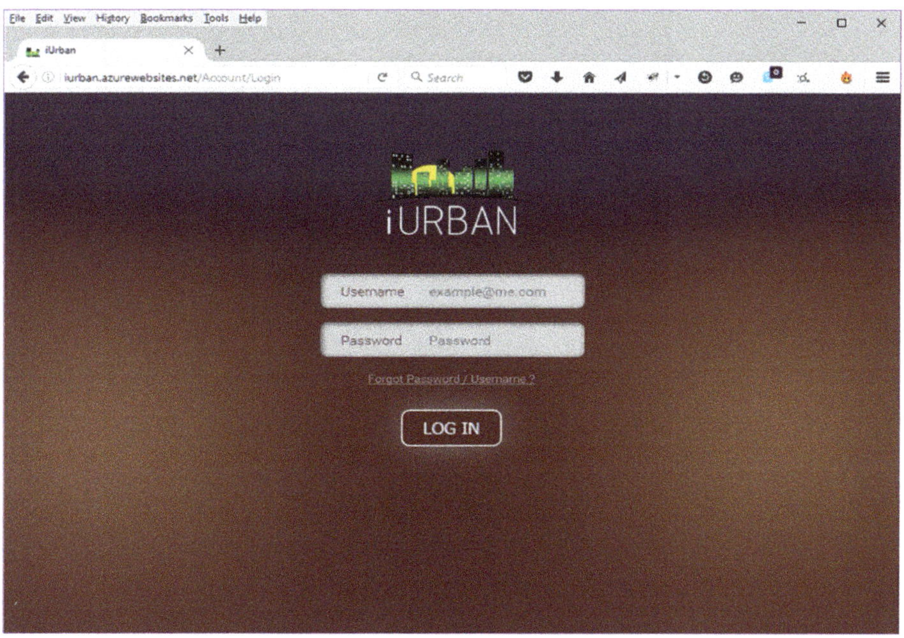

Figure 5.2 Access to LDSS GUI.

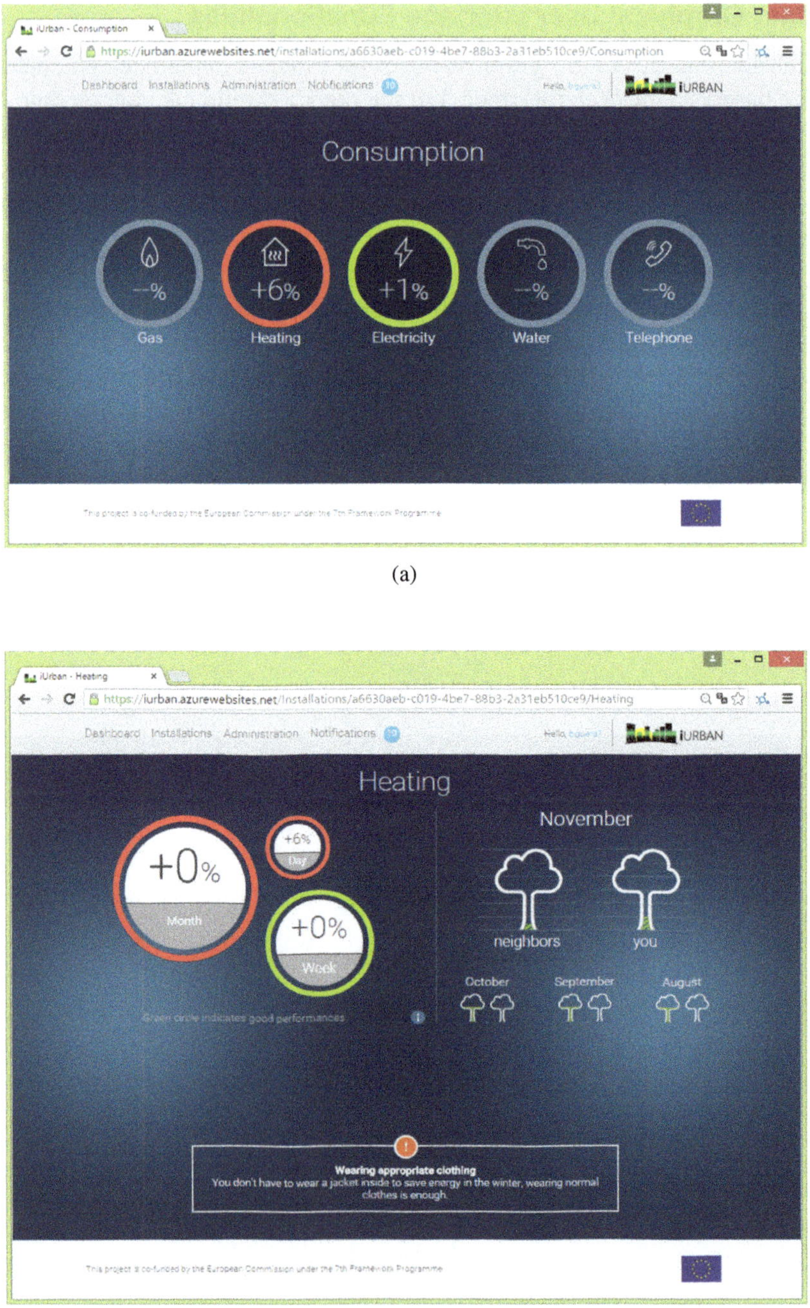

(a)

(b)

Figure 5.3 Energy visualization in iURBAN—(a) types and (b) default periods.

Concepts of user engagement have been introduced in the GUI, as shown in Figure 5.3(b), where energy tips and energy consumption efficiency based on trees achieved is shown.

Personalized messages can be sent every time the operator wishes, or can be scheduled over a period of time. Messages can be persistent or can be shown only during a few seconds. This provides a really powerful tool for communication between utility and their customers. Figure 5.4 shows an example.

The comparison between LDSS users will be performed using a concept called *Green Rewards*, represented with a tree. An average of the *Green Rewards* in all the installations included in the iURBAN pilots is performed to show LDSS end users how close are they with regard to the average (in the GUI named *neighbors*). User can know which day the tree was awarded by clicking a tree as shown in Figure 5.5.

Energy consumption and production predictions (for gas, water, and heating) have been also integrated into the user interface. This information is shown to users at the same level of real smart meter data as shown in Figure 5.6, where (a) shows the prediction of electricity consumption at the end of the day, while (b) and (c) provide the prediction of consumption by hours and days, respectively, superimposed with real consumption.

Users producing energy will be able to visualize the production in the same view as is shown energy consumption. So they will find production and consumption summary in a single view as shown in Figure 5.7.

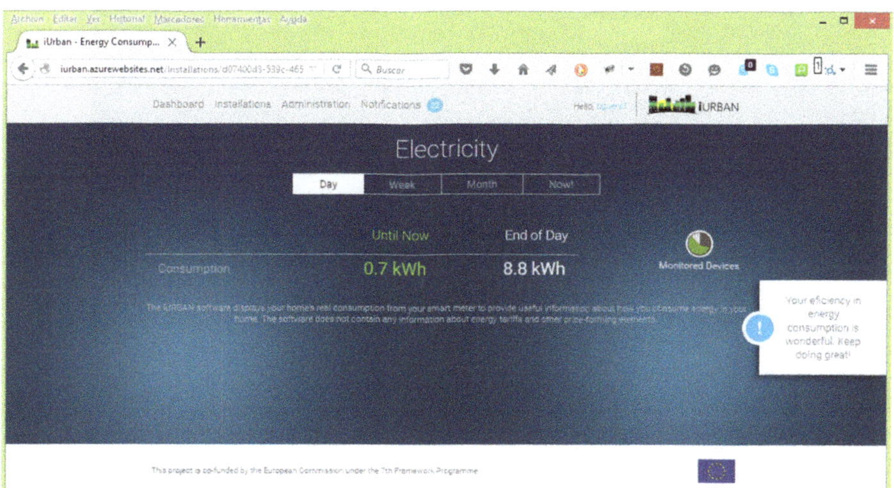

Figure 5.4 Personalized message.

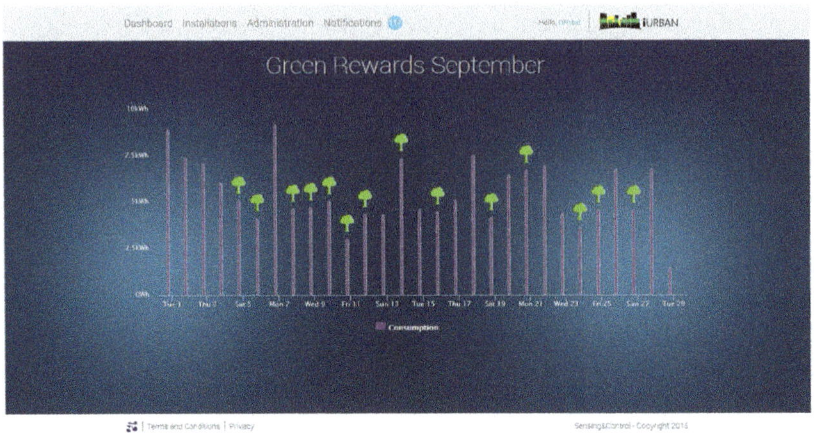

Figure 5.5 Tree award by day.

As previously commented, iURBAN deals with different types of energy besides electricity; Figure 5.8 shows screen captures for heating and water consumption.

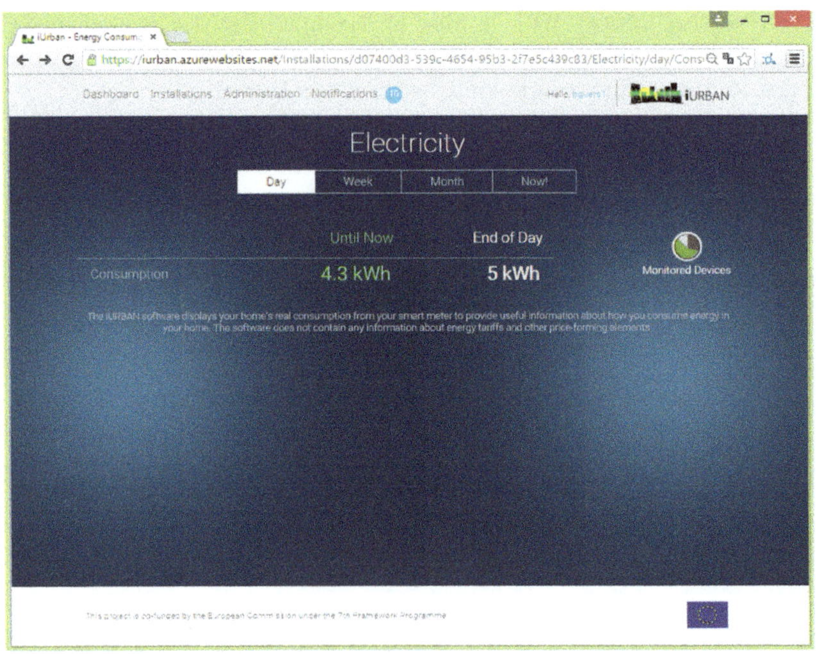

(a)

(b)

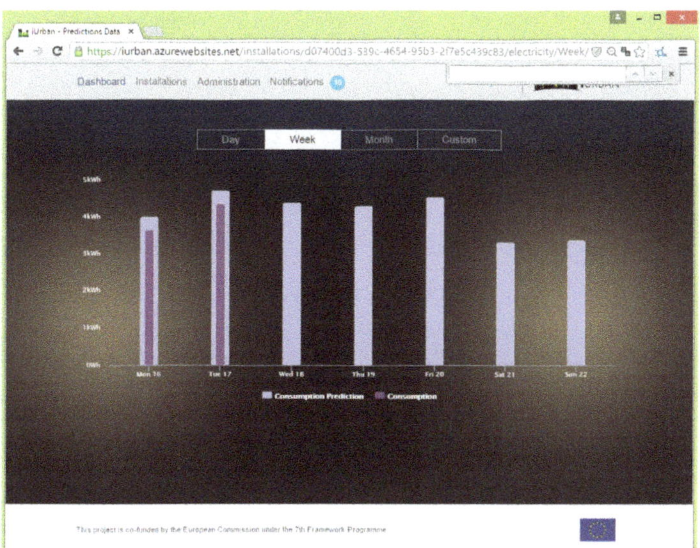

(c)

Figure 5.6 Energy visualization of energy consumption and energy prediction in iURBAN—(a) consumption up to given time of the day, and prediction for the end of the day, (b) detail of electricity consumption and prediction by hours, and (c) detail of electricity consumption and production by days.

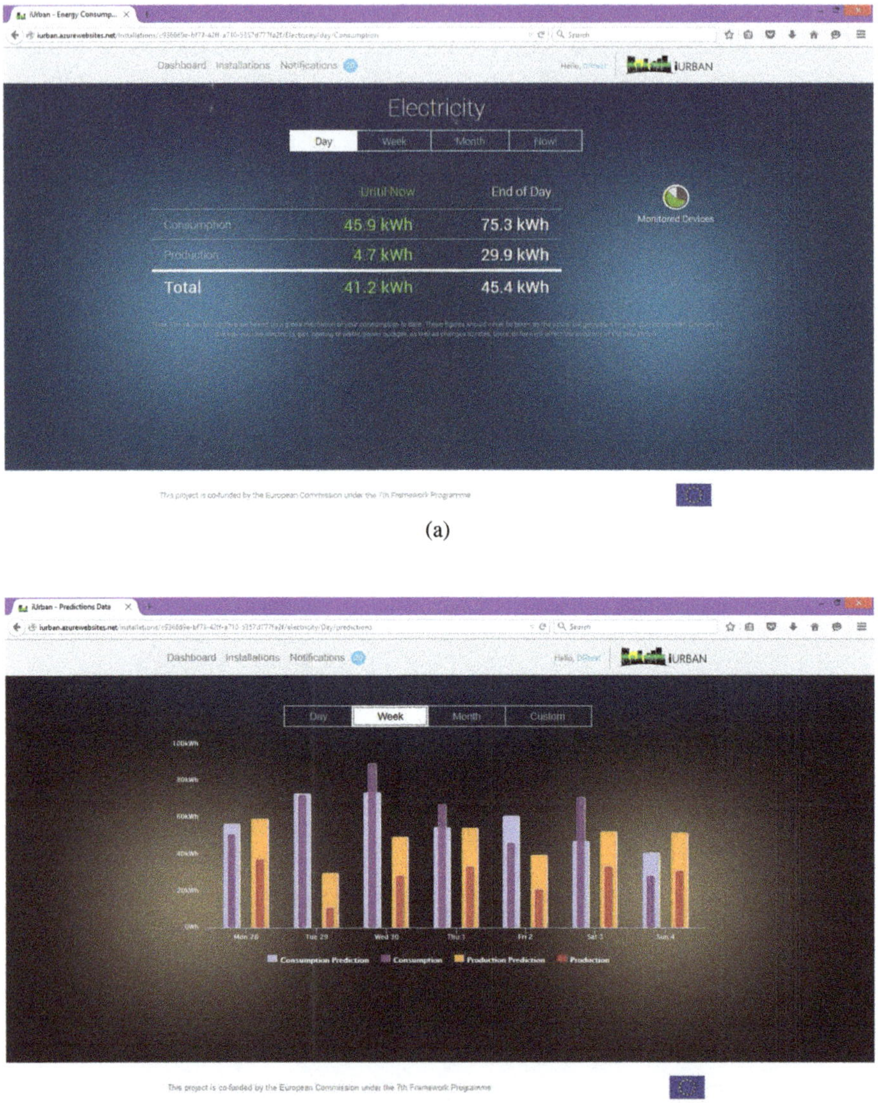

(a)

(b)

Figure 5.7 Energy visualization of energy consumption and energy prediction in iURBAN—(a) consumption and production summary, including predictions, (b) graphical view of week day.

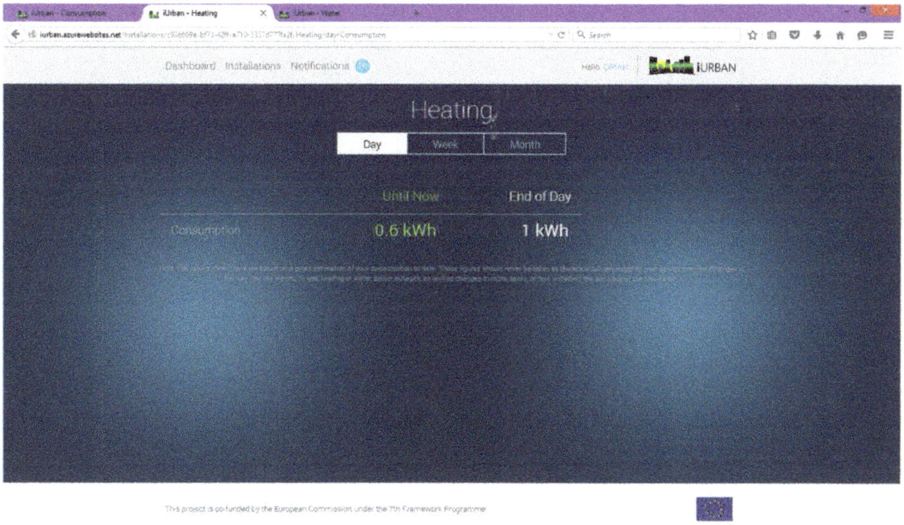

(a)

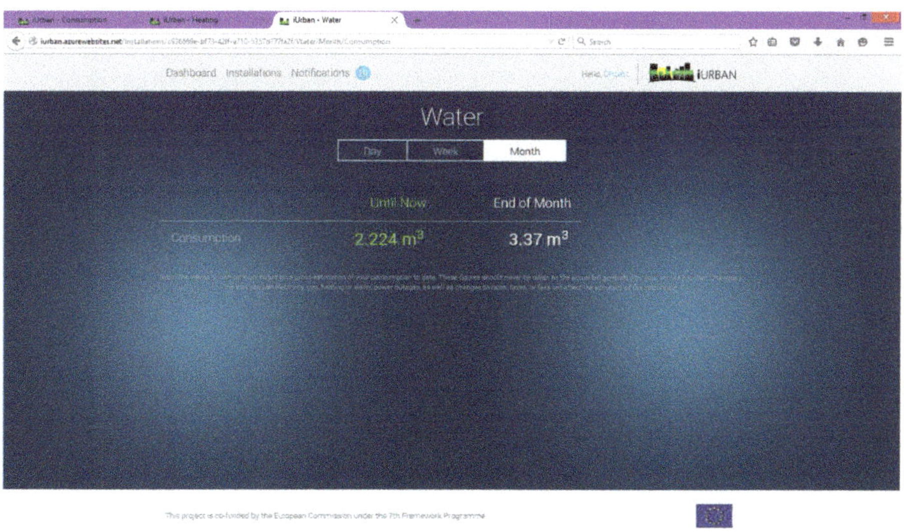

(b)

Figure 5.8 (a) Heating and (b) water consumption visualization in iURBAN.

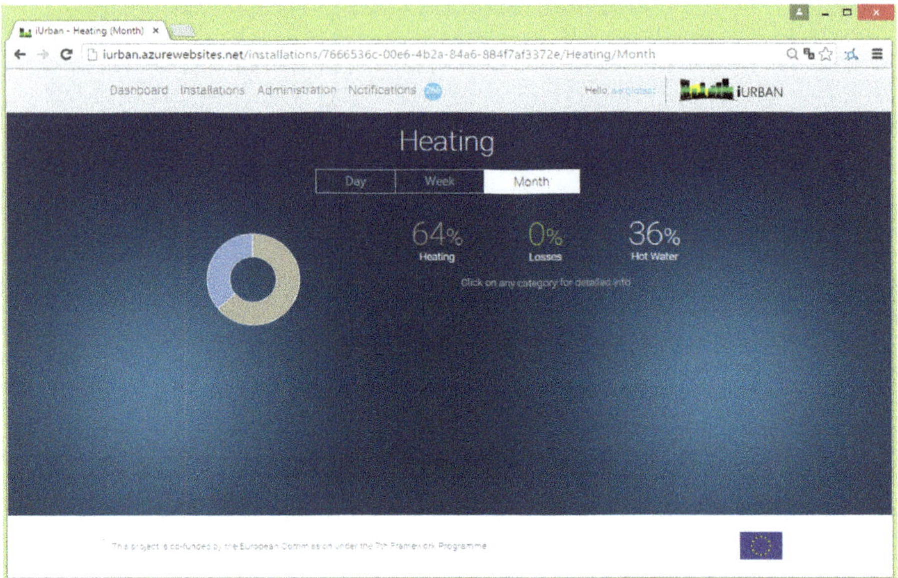

(a)

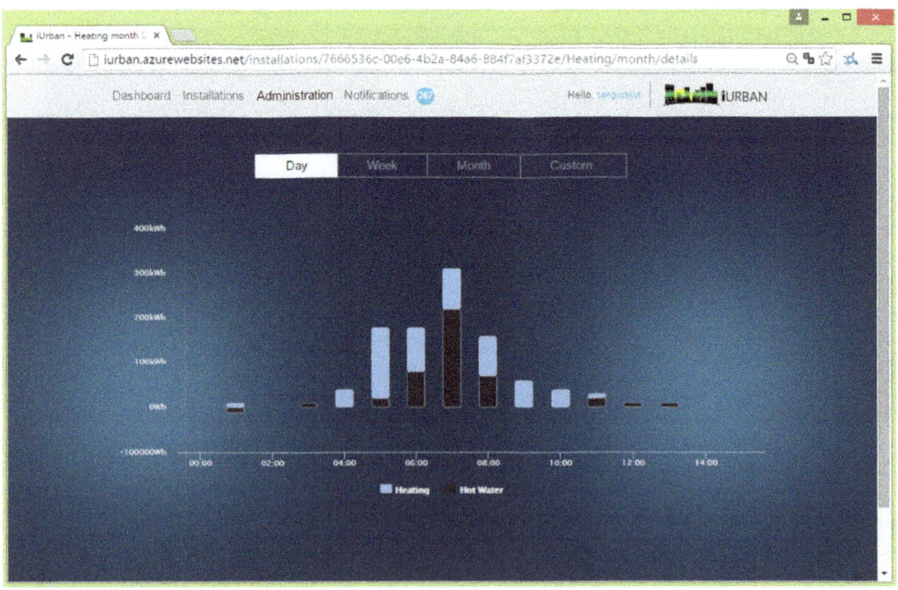

(b)

Figure 5.9 District heating energy consumption view.

In Plovdiv, some buildings use district heating as main source of heat to keep the living place warm during winter. For the households within these buildings, the LDSS has incorporated a representation of the energy consumed which is associated with district heating: (i) heat energy, (ii) hot water, and (iii) losses, in order to enable consumption analysis by the owners of the installations. Figure 5.9 provides an example of an installation.

The demand response has been introduced in the LDSS by means of two different views: (i) notifications (electricity and heating) and (ii) demand response action request.

The notification arrives each time CDSS sent a demand response action to a given installation. It provides information about the type of demand response and a link of the information related to it. Figure 5.10 shows demand response notification for the electricity and heating.

From notifications view, the user can get further information about the demand response requests. Figure 5.10 shows screen captures for the two types. While consumption DR is informative, the thermostat demand response action needs agreement by end user in order to allow LDSS to modify thermostat settings.

Thermostat DR is enabled through the smart home using Z-WAVE thermostats. Upon agreement from customer, the settings to the thermostat

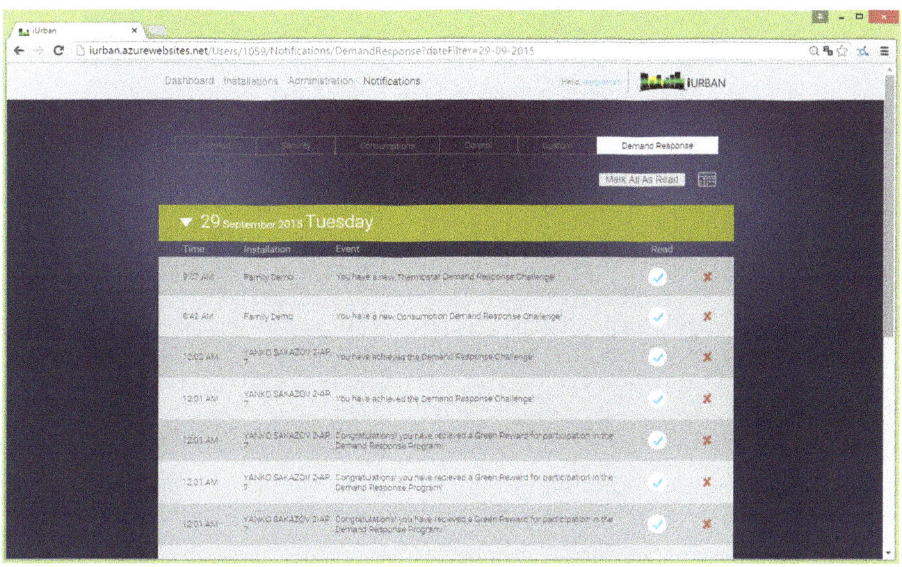

Figure 5.10 Demand response notifications view.

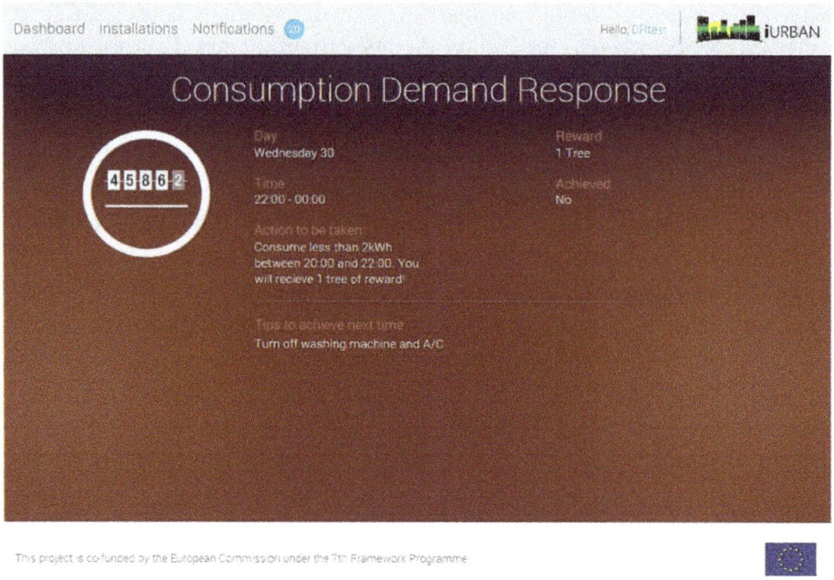

(a)

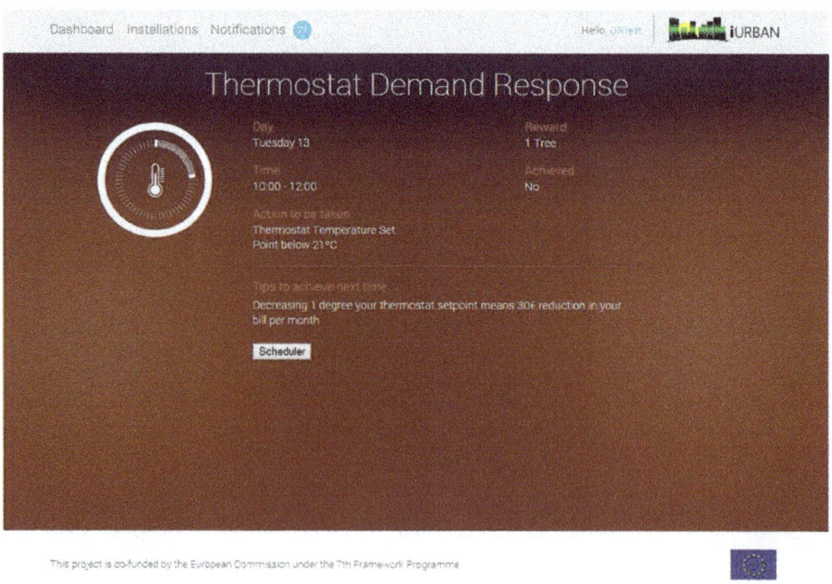

(b)

Figure 5.11 Demand response information for (a) consumption demand response and (b) thermostat demand response.

are set (i.e., fix thermostat target temperature) by means of scheduled action at a given time.

When a DR action is achieved, the user is informed by means of a notification. At any time, the user can review its achievements (coin icon in Figure 5.11), in comparison with the energy consumed and the trees awarded. The user can review the historical achievements as shown in Figure 5.12.

LDSS provides tools to end users to manage Z-wave devices at home. A graphical management interface has been developed in order to provide easy understanding process. Users are able to manage their installation, such as adding new devices to smart home, configuring them, and knowing its status just to provide few examples as shown in Figure 5.13.

As pointed out before, LDSS is also provided by means of Android and iOS app. The app comes with same functionalities as in the Web interface, with the exception of administration functions, which are not included. One of the main complements with respect to the Web interface is the real-time notifications by using the push channels of the smartphones, thus making more usable by the end users, especially for the purpose of demand response.

Figures 5.14 show a screen capture from iPhone.

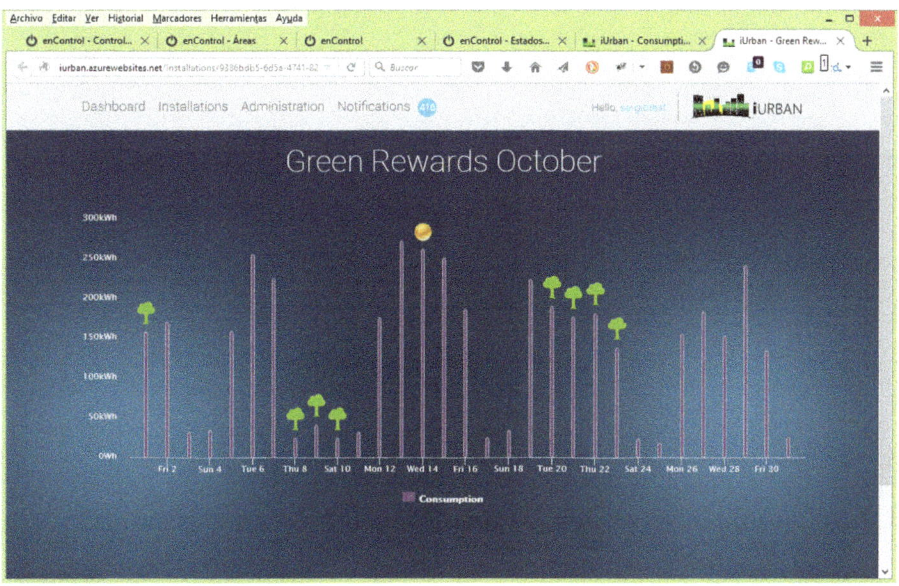

Figure 5.12 Demand response achievements.

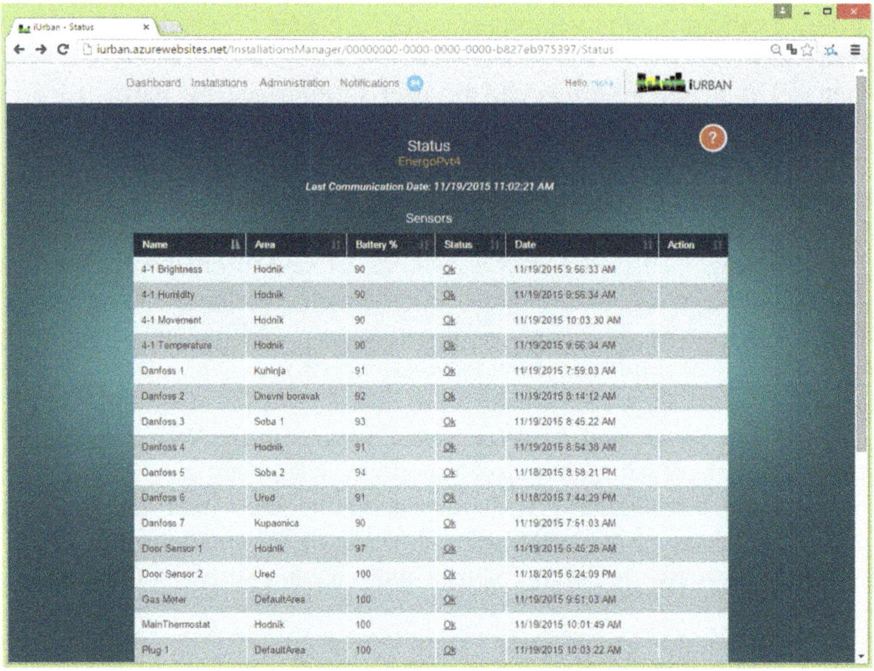

(a)

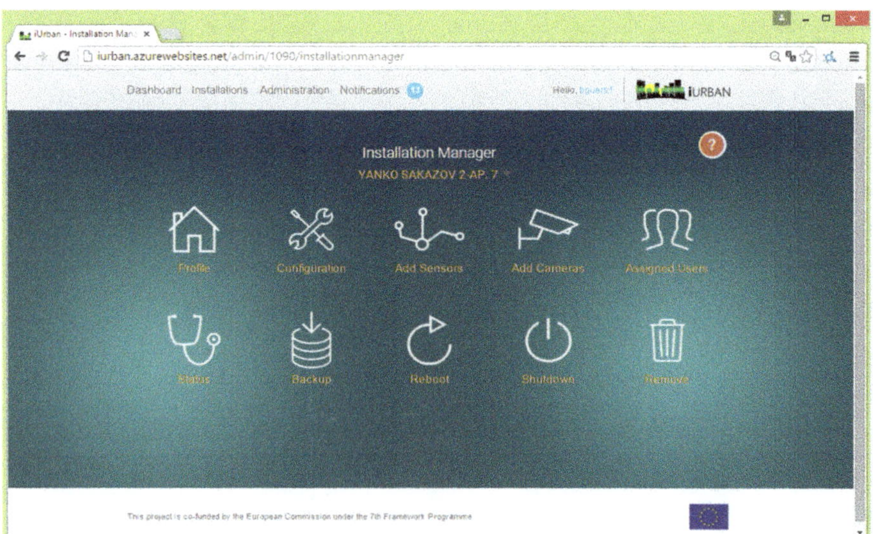

(b)

Figure 5.13 Z-wave devices management and status views—(b) general control, (a) device status, and (c) security device status.

(a)

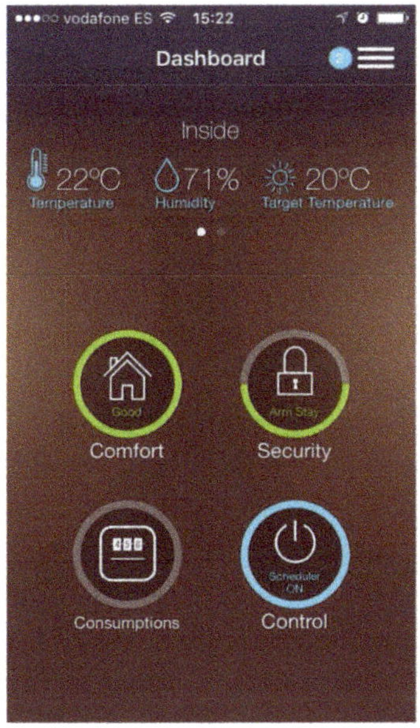

(b)

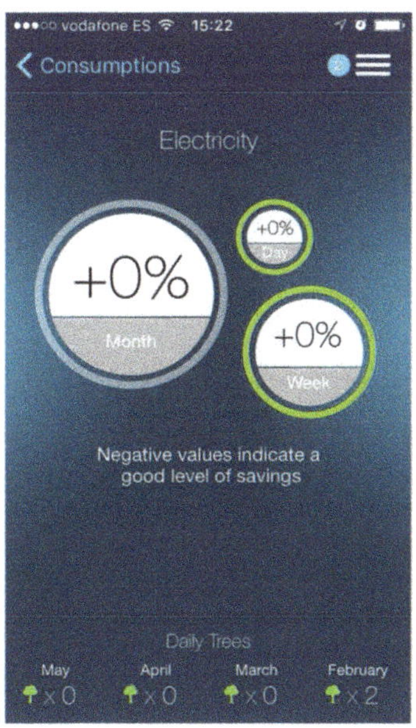

(c)

Figure 5.14 Captures from iPhone interface.

5.3 Conclusion

LDSS is a tool target to households and embraces visualization of energy data and engagement functions to foster knowledge on the way energy is consumed and promote energy aware habits.

It has been designed with the help of end users, as they have taken active part during the development of the graphical user interface. We believe this exercise increased the chances to succeed in the acceptance of the final solution by end users, particularly to understand what it is important for them and what is not.

LDSS has been designed as Web interface, but Android and iOS apps have also been developed; in particular, the push notifications offered by the smartphones were extremely useful to drive demand response programs and communication with users more friendly.

The combination of energy information (from the utilities) and the capabilities of smart home of the LDSS have been reported as key for end users to install in their homes, especially the security and remote control of home heating system.

6

Virtual Power Plant

Mike Oates and Aidan Melia

Integrated Environmental Solutions, Glasgow, UK

Abstract

In the context of this chapter the virtual power plant (VPP) is considered as a high level design tool based upon load aggregation of near real-time metered energy demand and generation data at building/apartment levels.

Target users, city planners and utility companies, will be able to use the VPP to gain an understanding of energy demand/generation at user defined and selected levels of interest ranging from high level city planning to the selection of individual buildings or user defined energy networks and so on. 'What if' scenarios aid in future development and planning of cities.

This chapter outlines development and integration of the VPP within the iURBAN ICT architecture. Content is also provided on the VPP; graphical user interface, calculation engine and application via case studies taken from the iURBAN project.

Keywords: Virtual power plant, Central decision support system, iURBAN ICT architecture, City model, High level city planning, Distributed energy resources.

6.1 Introduction

This chapter outlines the development work required for the virtual power plant (VPP) as part of the iURBAN [1] project.

What is a VPP?

There are numerous definitions for a virtual power plant. Below are a few examples:

> *"a VPP is a system that relies upon software and a smart grid to remotely and automatically dispatch and optimize Distributed Energy Resources (DER)s via an aggregation and optimization platform linking retail to wholesale markets."* [2]

> *"a VPP as a system that integrates several types of power sources, (such as micro combined, heat and power (CHP), wind-turbines, small hydro, photovoltaics (PV), back-up generators, batteries etc.) so as to give a reliable overall power supply."* [3]

Further definitions can be found at [4, 5].

It is reported that due to increased activity in smart meter installations and other smart grid technologies, as well as challenges in balancing variable renewable generation on the grid, it is reported that total annual VPP vendor revenue will grow from $1.1 billion in 2014 to $5.3 billion in 2023 [6].

6.2 Virtual Power Plant in iURBAN

In the context of the iURBAN project, it was agreed that the VPP should not follow that of detailed network modeling software available on the current market, such as GridLAB-D^{TM} [7]. The VPP is developed as a high-level design tool. The modeling approach of the VPP is based upon load aggregation of near real-time metered energy demand and generation data and modeling of electricity and heat generation at building/apartment and district level. City planners and utility companies will be able to undertake VPP analysis to gain an understanding of energy demand and generation and the associated costs at selected levels of interest, ranging from high-level city planning to the selection of individual buildings or user-defined energy networks.

Figure 6.1 gives a simplified overview of the iURBAN information and communications technology (ICT) architecture. The broader overview of Figure 6.1 iURBAN ICT architecture includes components that combine to form the SMART urban decision support system (smartDSS) developed by the project.

6.2.1 smartDSS

The smartDSS consists of the following components (some of which have been described in other chapters):

- Local decision support system graphical user interface (LDSS GUI),
- LDSS part of the Smart City Database (SCDB-LDSS),

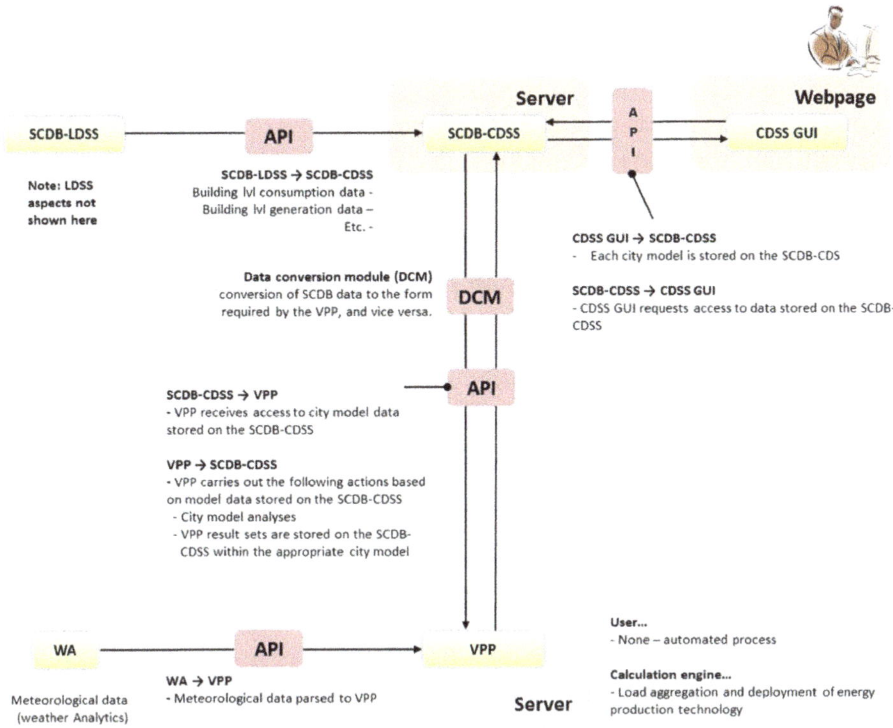

Figure 6.1 iURBAN ICT architecture—CDSS/VPP relationship diagram.

- Smart City Prediction Algorithms (SCPA),
- Meteorological data parsed from weather analytics (WA) [8].
- Central decision support system graphical user interface (CDSS GUI).
- CDSS part of the Smart City Database (SCDB-CDSS).
- VPP.

6.2.2 LDSS

A brief summary of the local decision support system (LDSS) is included for clarity. Within iURBAN, the LDSS GUI is a tool used by occupants of private buildings, apartments, and public municipality buildings such as kindergartens (schools), offices, and leisure centers.

Demand and generation metered data from buildings within the iURBAN demonstrations cities is parsed to the LDSS part of the SCDB-LDSS. These data along with meteorological data from WA are parsed to the SCPA. The SCPA component performs analysis on the data and generates forecast demand

and generation data up to 72 hours ahead, which is stored in the SCDB-LDSS. The LDSS GUI primary focus is to educate users on how and where energy is being consumed with a view to encouraging users to make savings, energy, and monetary, and reduce greenhouse gas emissions. The LDSS GUI also includes demand response (DR) actions.

6.2.3 CDSS

A brief summary of the central decision support system (CDSS) is included for clarity. The CDSS GUI is a tool used by utilities and municipalities. The CDSS GUI enables users to view metered data at a city scale and focused areas of interest such as district and neighborhood level, to make informed decisions on high-level planning.

There is an application program interface (API) layer that exists between the SCDB-LDSS and the SCDB-CDSS. This API parses stored metered and forecast data from the SCDB-LDSS to the SCDB-CDSS. The SCDB-CDSS and CDSS-GUI are both developed by the iURBAN partner Vitrociset [9].

6.2.4 VPP

The VPP, developed by IES [10], is a back end calculation engine to the CDSS GUI which acts as a front end to the VPP. The VPP is parsed, via an API layer, city model data stored on the CDSS part of the Smart City Database (SCDB-CDSS). The VPP writes the results of the calculations back to the SCDB-CDSS for access by the CDSS GUI.

6.3 User Interface

As discussed, the CDSS GUI acts as the front end to the VPP calculation engine. Figures 6.2–6.6 are screenshots of the CDSS GUI focusing on VPP functionality.

City models (explanation given within the following section) are created by CDSS GUI users. Figure 6.2 gives an example of saved city models stored on the SCDB-CDSS.

CDSS GUI users can create a new city model by clicking on the "New City Model" button, as shown in Figure 6.2. Upon clicking the button, a blank "City Model" screen appears, refer to Figure 6.3. Users can create city models for both Plovdiv and Rijeka. As explained within the below section, users define electricity and heating networks by creating network models consisting of site data, commodities, buildings, nodes, and installations.

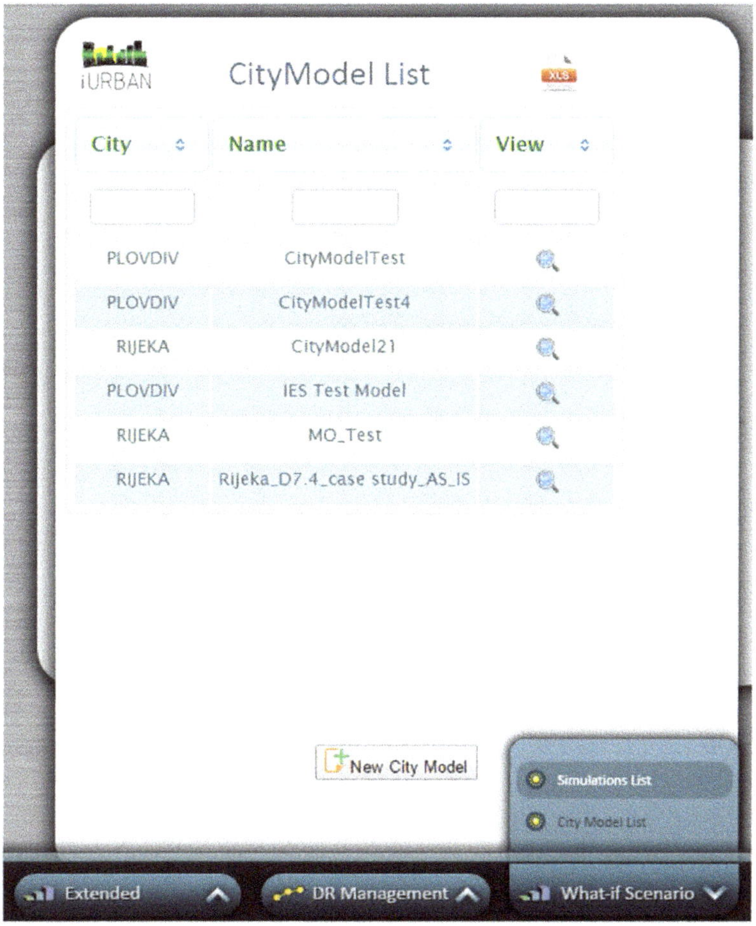

Figure 6.2 CDSS GUI—city model list.

An example of a simple city model is given within Figure 6.4 and consists of the following:

- Name: CityModel2
- City: Rijeka
- Site data: Lat. = 40, Long. = 13
- Commodities: Electricity commodity account
- Buildings: Buildings 1–4 (names removed)
- Nodes: Electricity root node (ELE1)
- Installations: None, baseline "as is" model.

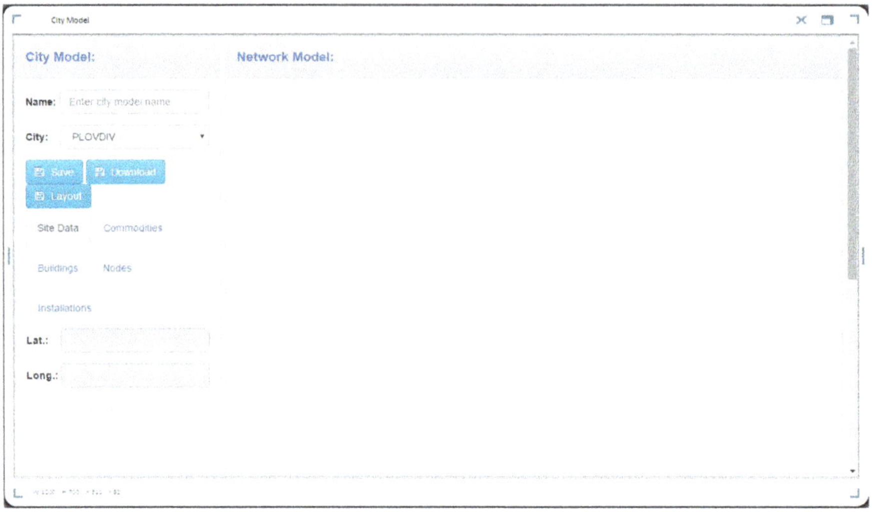

Figure 6.3 CDSS GUI—create city model.

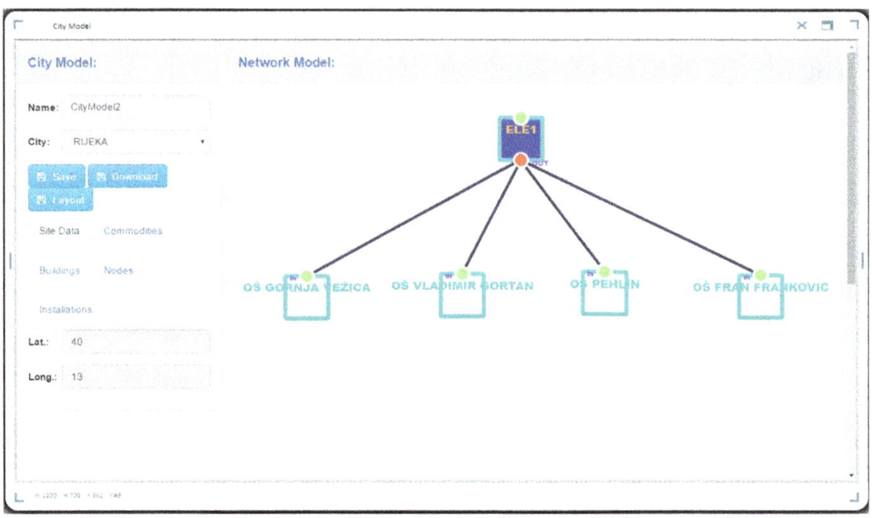

Figure 6.4 CDSS GUI—example city model.

Figure 6.4 represents an "as is" city model, refer to explanation given within the following section.

Figure 6.5 illustrates parameters required to enact a VPP simulation run. Users select the city model of interest, type of analysis, start/end dates, step,

Figure 6.5 CDSS GUI—VPP run settings.

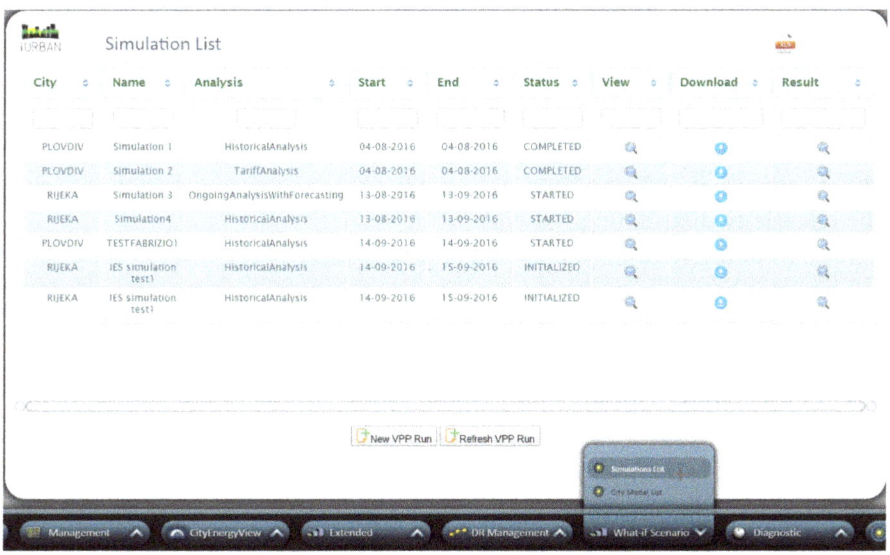

Figure 6.6 CDSS GUI—VPP simulation list.

and time interval. XML files, consisting of city model and VPP simulation run parameters, are parsed to the VPP engine. Metered data resides behind selected buildings of interest and are a driver for VPP calculations.

An example list of VPP simulation runs is shown in Figure 6.6. The CDSS GUI provides a status update. Results can be viewed or downloaded for further analysis.

The above process is repeated to create "what if" variants of the "as is" city model, refer explanation given within the following section.

6.4 City Models

The CDSS GUI enables users to create different models for different purposes such as modeling regions of a city, types of building, energy supply and management technologies, or degrees of modeling detail appropriate to particular tasks.

In addition, city models managed by the CDSS GUI user are conceptually divided into two categories:

- "as is"—city models representing the structures and consumption patterns currently in place (the status quo).
- "what if"—city models representing possible alternatives.

Examples of "what if" variant city model(s) include the addition of DER, electricity storage, modified demand from buildings, and electric vehicles. "What if" variant model(s) can answer questions such as the following:

- What is the likely effect of adding PV arrays to certain buildings?
- What is the likely effect of adding electricity storage at a certain point in the electricity distribution network?
- What is the likely effect of introducing a district CHP plant to serve a certain area?
- What is the likely effect of introducing a large-scale PV farm to serve a certain area?
- What is the likely effect of introducing tariffs in monetary and energy consumption terms?

The "as is" and "what if" city models allow for cross comparisons to be made between models.

6.5 Modeling Approach

City model data parsed to the VPP from the CDSS part of the SCDB-CDSS is referred to as the VPP city model. The VPP city model is formed around the following features: commodity, external supply, commodity account, fuel, carbon dioxide (CO_2) emissions, distribution network, network node, transmission channel, prosumer object, DER object, generator, storage device, and manager, which are illustrated in Figure 6.7. The figure illustrates an electricity distribution network defined for a city model. In this diagram, rectangles with dashed borders represent CDSS GUI objects, which serve as receptacles for VPP objects. VPP objects fall into three basic categories: nodes (colored discs), DER objects (color-filled rectangles), and managers (color-filled diamonds).

DER objects are further categorized according to type such as the following:

- Electricity network;
  - Power station,
  - CHP,
  - PV array,
  - Wind turbine,
  - Electrical storage.
- Heat network;
  - Heat generator,
  - Electric heat pump,
  - CHP,
  - Solar water heating,
  - Thermal storage.

The filled rectangles at the base of the diagram are prosumer units by means of which demands represented by input time series are connected to the system. Other VPP objects function algorithmically as a function of other variables in the model.

6.6 Case Study: Rijeka, Croatia

Figure 6.8 is a CDSS GUI demonstration of the iURBAN metered installations within Rijeka, Croatia, were 3 public kindergartens, 4 public schools, 12 residential, 4 sports centers, 2 culture centers, and 7 heating plant installations. CDSS GUI legend notation is shown in Figure 6.9.

6.6.1 "As is" Scenario

Due to the sparse nature of building locations, limited network topology information and metered installations within the framework of the iURBAN

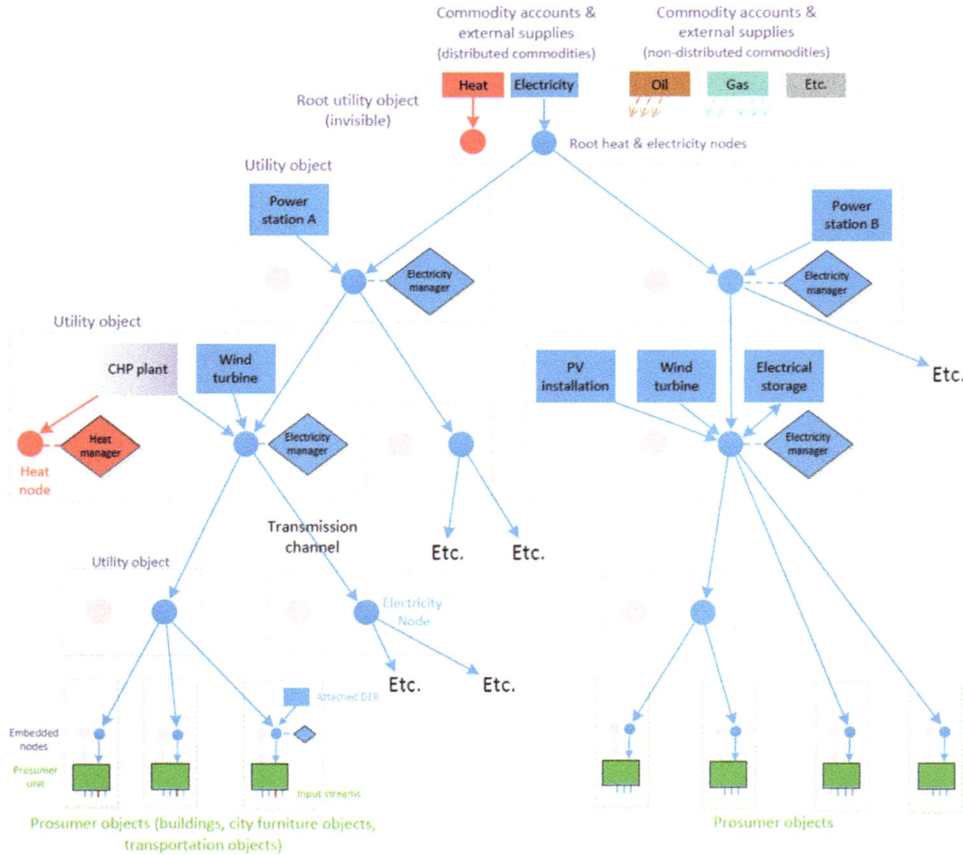

Figure 6.7 Part of an example city model focusing on the electricity distribution network.

Figure 6.8 CDSS GUI—Rijeka, Croatia. Red dashed circles denote defined building clusters.

Figure 6.9 CDSS GUI legend.

project the Rijeka case study have been simplified. Figure 6.10 illustrates a simple VPP electricity network model for Rijeka.

In reference to the VPP modeling approach in Figure 6.7, five electricity sub-stations have been formed, representative of utility objects. Buildings within close proximity of a sub-station have been grouped together, refer to Figures 6.9 and 9.10. Buildings in this context are representative of prosumer objects in Figure 6.7.

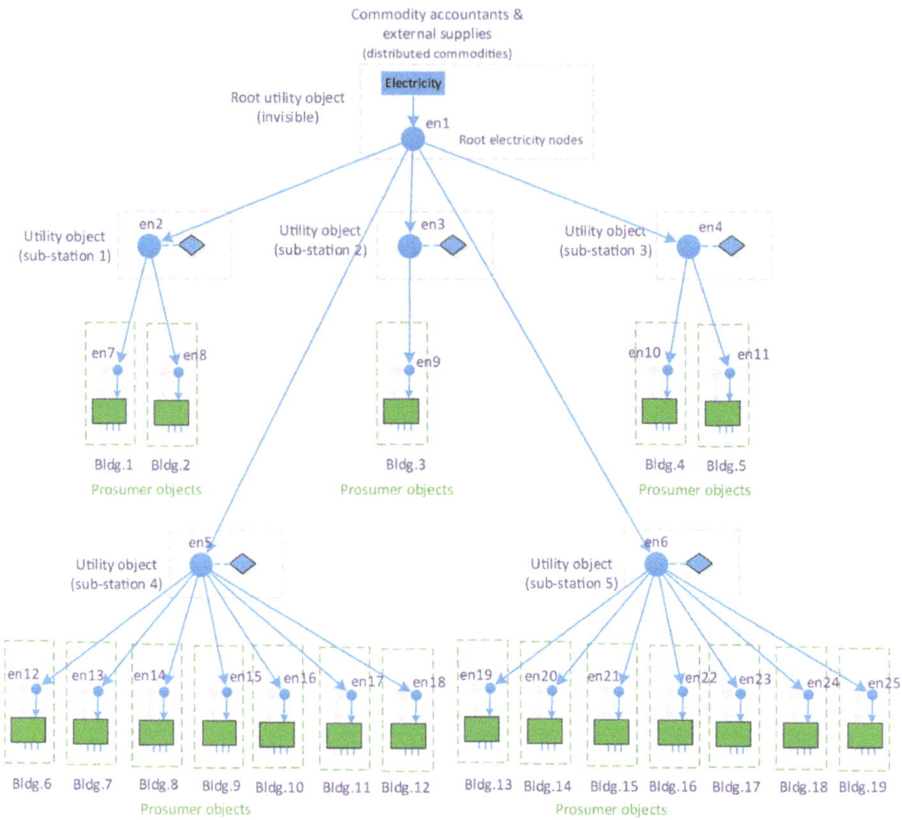

Figure 6.10 Simple VPP electricity network model—Rijeka, Croatia.

Further to sensitivity analysis carried out on the SCDB-CDSS data, the below period of analysis is considered for selected meter installations. The period of analysis was selected on the criteria of having the maximum number of meters with non-missing data for a whole week.

Data summary:

- Location: Rijeka (CityModelName calls weather analytics (WA) API for Rijeka weather data).
- Latitude: N/A (WA API).
- Altitude: N/A (WA API).
- Scenario: As is—electricity network.
- Start date/time: 2016-03-01 [00:00].
- End date/time: 2016-03-07 [23:00].

- Reporting period: Hourly (on the hour).
- Electricity installations devised into five hypothetical electricity sub-stations, refer to Figures 6.9 and 6.10.

The iURBAN project does not have access to all utilities, such as electricity, district heating, gas, and water, meter readings for each installation. Table 6.1 lists building types used within Rijeka case study and denotes available electricity consumption and production meters. Each building type is grouped to a hypothetical electricity sub-station.

Upon finalization of the "as is" city model within the CDSS GUI, the following extensible markup language (XML) files are created and parsed to the VPP engine via the CDSS part of the SCDB-CDSS:

- CityModel.xml—i.e., electricity network topology.
- Commodities.xml—i.e., commodity attributes such as CO_2 emission factors.
- Run.xml—i.e., VPP run specifier settings such as start/end date and time.
- Timeseries.xml—i.e., time series meter installation readings.

Table 6.1 Hypothetical electricity sub-stations 1–5

Hypothetical Sub-stations	Prosumer Object Name	Type	Meter (Consumption)	Meter (Production)
1	Building 1	Public kindergarten	Yes	Yes
"	Building 2	Sports center	Yes	Yes
2	Building 3	Public school	Yes	No
3	Building 4	Public school	Yes	Yes
"	Building 5	Public kindergarten	Yes	Yes
4	Building 6	Residential	Yes	No
"	Building 7	Sports center	Yes	No
"	Building 8	Residential	Yes	No
"	Building 9	Residential	Yes	No
"	Building 10	Residential	Yes	Yes
"	Building 11	Culture center	Yes	No
"	Building 12	Culture center	Yes	No
5	Building 13	Residential	Yes	No
"	Building 14	Sports center	Yes	No
"	Building 15	Public school	Yes	No
"	Building 16	Residential	Yes	No
"	Building 17	Residential	Yes	No
"	Building 18	Public kindergarten	Yes	No
"	Building 19	Public school	Yes	No

Source: Author.

6.6.2 "What if"—Scenarios

The below "what if" city models are an example of the application of creating variant models for comparison against the "as is" model.

Table 6.2 outlines selected DER installations for prosumer objects (buildings 1–19) and utility objects (sub-stations 1–5) across 8 scenarios. For example, scenario 1 includes a 1No. 1 kW (small domestic) photovoltaic (PV) panel assigned individually to all 19 buildings; sub-stations have not been assigned DER installations. Scenario 2 is the same as scenario 1 with the exception that the capacity rating of the PV panel installation is changed from 1 kW to 6 kW (large domestic). Wind turbines are modeled in scenarios 3 and 4. Scenarios 5–8 include energy storage device in addition to PV and wind turbines for selected scenarios. For example, scenario 5 includes 1No. 1 kW PV panel and 1No. 50 kWh energy storage device assigned individually to all 19 buildings and so on.

DER installation parameters are shown in Tables 6.3–6.5; installation types refer to the number shown in Table 6.2.

6.6.3 Results

Rijeka case study VPP results for "as is" and "what if" scenarios are shown in Table 6.6. Results are presented for the electricity commodity account, i.e., root node, not for each electricity node in the network.

For scenarios 1–4, no electrical energy storage devices, the results reflect the expected behavior of the electricity network when compared against the "as is" scenario, i.e., the baseline scenario. The introduction of renewables, PVs, and wind turbines reduces demand from external supply to the network. This is a result of renewables offsetting electricity demand from buildings 1 to 19. CO_2 emissions are also reduced within the model based on the reduction in fossil fuel-based external supply to the network. There is a noticeable difference between the PV and wind turbine scenario results. This difference is due to the variation in wind speed for Rijeka based on its coastal location; average, maximum, and minimum wind speed for the selected period of analysis is 3.44 m/s, 10.19 m/s, and 0.89 m/s. A wide range of wind turbine rated power, 1.5–15 kW per installation, is modeled. This may be considered unrealistic, but the model gives an indication of future potential if such solutions were considered feasible to install.

In scenarios 5–8, which include electrical energy storage in addition to selected renewable technology, external supply and carbon emission reduce when compared to scenarios 1–4 with no electricity storage installations.

Table 6.2 Scenario parameters

	Building 1	Building 2	Building 3, 4, ..., 16, 17	Building 18	Building 19	Sub-station 1	Sub-station 2	Sub-station 3	Sub-station 4	Sub-station 5
Scenario 1										
PV (kW)	1	1	1	1	1	–	–	–	–	–
Scenario 2										
PV (kW)	6	6	6	6	6	–	–	–	–	–
Scenario 3										
Wind (kW)	1.5	1.5	1.5	1.5	1.5	–	–	–	–	–
Scenario 4										
Wind (kW)	15	15	15	15	15	–	–	–	–	–
Scenario 5										
PV (kW)	1	1	1	1	1	–	–	–	–	–
Energy storage (kWh)	50	50	50	50	50	–	–	–	–	–
Scenario 6										
PV (kW)	6	6	6	6	6	–	–	–	–	–
Energy storage (kWh)	50	50	50	50	50	–	–	–	–	–
Scenario 7										
Wind (kW)	1.5	1.5	1.5	1.5	1.5	–	–	–	–	–
Energy storage (kWh)	50	50	50	50	50	–	–	–	–	–
Scenario 8										
Wind (kW)	15	15	15	15	15	–	–	–	–	–
Energy storage (kWh)	50	50	50	50	50	–	–	–	–	–

Source: Author.

Table 6.3 PV array parameters

Installation Category	Area (m^2)	Azimuth (Clockwise from North)	Inclination (from Horizontal)	PV Module Nominal Efficiency	Nominal Cell Temperature (NOCT) (°C)
PV 1 kW (small domestic)	7.2	180	35	0.1100	45.0
PV 6 kW (large domestic)	43.2	180	35	0.1100	45.0

Reference Irradiance for NOCT (W/m2)	Temp. Coefficient for Module Efficiency (1/K)	Degradation Factor	Shading Factor	Electrical Conversion
800	0.0040	0.99	1.0	0.85
800	0.0040	0.99	1.0	0.85

Table 6.4 Wind turbine parameters

Installation Category	Rated Power (kW)	Hub Height (m)	Power Curve [wind speed (m/s), Power Output Fraction %]
Wind 1.5 kW (house)	1.5	5	0 0 4 0.1 7 0.5 12 0.8 25 1
Wind 15 kW (farm)	15	5	0 0 4 0.1 7 0.5 12 0.8 25 1

Table 6.5 Electrical energy storage parameters

Installation Category	Storage Capacity (kWh)	Initial Storage Energy (kWh)	Storage Method	Losses
lithium-ion Battery 50 kWh	50	2	lithium-ion Battery	0

Table 6.6 Rijeka VPP results, percentage difference against "as is" baseline model

Scenario	External Supply (% Diff)	External Indirect CO_2 Emission (% Diff)
As is (baseline model)	–	–
Scenario 1—1 kW PV	0.30	0.30
Scenario 2—6 kW PV	1.75	1.67
Scenario 3—1.5 kW wind turbine	0.78	0.77
Scenario 4—15 kW wind turbine	7.40	7.30
Scenario 5—1 kW PV and 50 kWh storage	0.31	0.31
Scenario 6—6 kW PV and 50 kWh storage	1.45	1.39
Scenario 7—1.5 kW wind turbine and 50 kWh storage	0.73	0.73
Scenario 8—15 kW wind turbine and 50 kWh storage	5.64	5.58

This is a result of model setup where electricity storage devices have been applied at building level. In some cases, on-site electricity generation from renewables at building level exceeds electricity building level demands. In this case, excess electricity generation charges on-site storage prior to being fed back to the electricity network upon electricity storage equaling storage capacity. In other instances, electricity demand at building level exceeds on-site electricity generation. This results in an electricity residual demand from the electricity network, sub-station, and then parent node, which in turn equates to an increase in external supply and CO_2 emissions to the model. This could be adverted with better electricity storage controls and storage at sub-station level. Future scenarios will look to address this.

6.7 Future Work

Future work consists of refining the Rijeka case study based on detailed electricity network topology. Other work includes the modeling of the second iURBAN case study in Plovdiv, Bulgaria.

Future VPP development includes the following:

- District cooling
- Water networks
- Optimized control strategies.

6.8 Conclusion

This chapter has provided information about the approach implemented within iURBAN with respect to virtual power plant development. This chapter also presents a use case example for Rijeka in Croatia, one of the iURBAN demonstration cities.

The VPP is developed as a high-level design tool. The modeling approach of the VPP is based upon load aggregation of near real-time metered energy demand and generation data and modeling of electricity and heat generation at building/apartment and district level.

Through use of the CDSS GUI, city planners and utility companies will be able to undertake VPP analysis to gain an understanding of energy demand and generation and the associated costs at selected levels of interest, ranging from high-level city planning to the selection of individual buildings or user-defined energy networks.

References

[1] iURBAN. (2016). iURBAN. Available at: http://www.iURBAN-project.eu/. Accessed 24 May 2016.

[2] Asmus, P. (2014). How real are virtual power plants? Available at: http://www.elp.com/articles/powergrid_international/print/volume-19/issue-11/features/how-real-are-virtual-power-plants.html. Accessed 24 May 2016.

[3] Landsbergen, P. (2009). *Feasibility, beneficiality, and institutional compatibility of a micro-CHP virtual power plant in the Netherlands. Master Thesis.* Published by TU Delft. http://repository.tudelft.nl/view/ir/uuid%3Aee01fc77-2d91-43bb-83d3-847e787494af/

[4] Greenergy. (2016). *The definition of the virtual power plant.* [online] Greenergy.hu. Available at: http://www.greenergy.hu/vpp_e.html. Accessed 18 Jul. 2016.

[5] Zurborg, A. (2016). Unlocking customer value: The virtual power plant, Department of Energy. Available at: http://energy.gov/oe/downloads/unlocking-customer-value-virtual-power-plant. Accessed 18 Jul. 2016.

[6] Navigant Research. (2014). Virtual power plants. Available at: http://www.navigantresearch.com/research/virtual-power-plants. Accessed 24 May 2016.

[7] GridLAB-D. (2012). GridLAB-D. Available at: http://www.gridlabd.org/. Accessed 24 May 2016.

[8] Weather Analytics. (2016). Weather analytics. Available at: http://www.weatheranalytics.com/wa/. Accessed 24 May 2016.

[9] Vitrociset. (2016). Vitrociset. Available at: http://www.vitrociset.it/index.php?lang=en. Accessed 24 May 2016.

[10] IES Ltd. 2016. Integrated environmental solutions. Available at: https://www.iesve.com/. Accessed 24 May 2016.

7

iURBAN Smart Algorithms

Sergio Jurado and Alberto Fernandez

Sensing & Control, Barcelona, Spain

Abstract

This chapter presents the smart algorithms developed in iURBAN. Prediction of consumption and production of energy, dynamic tariff comparison and demand response.

Keywords: Energy prognosis, Demand response, Dynamic tariff.

7.1 Introduction

This chapter outlines details for the following smart algorithm modules:

- "As is" generation and consumption forecasts using artificial intelligence (AI).
- Dynamic tariff comparison and demand response (DR) simulations.

The algorithms for DR analysis, renewable energy production analysis, and tariff analysis were developed with respect to the electricity grid. The algorithms are integrated with the virtual power plant (VPP) to perform energy analysis of the overall city in order to enable balancing energy at both local and centralized levels.

Based on forecasting weather data which is be stored in the Smart City Database (SCDB), the prediction algorithms are able to predict the energy consumption, production, transfer, and storage of energy at the city level. For example, for the next 24, 48, and 72 hours, they can advice how to reduce consumption (covered by the local decision support system.), how to integrate new technologies, how to optimize existing technologies, and how to determine the optimum time to buy and sell energy to the electricity and/or vehicle grids or to store energy from centralized DER sources and assist iURBAN decision support systems.

The prediction algorithms consist of 3 modules: the AI module, the tariff analysis module (including DR aspects), and the weather forecast module (weather data are obtained from professional weather data provider www.weatheranalytics.com).

7.2 "As is" Generation and Consumption Forecasts

7.2.1 Introduction

A large variety of AI techniques have been applied in the field of short-term electricity consumption forecasting, showing a better performance than classical techniques. Specifically, machine learning has been proven to accurately predict electric consumption under uncertainties. For instance, Khamis proposes in [1] a multilayer perceptron neural network to predict the electricity consumption for a small-scale power system, obtaining a better performance than with traditional methods, while Marvuglia et al. consider Elman neural network for the short forecasting of the household electric consumption with prediction errors under 5% [2]. Also in [3], a study of electric load forecasting is carried out with classification and regression trees (CART) and other soft computing techniques obtaining again better results than classical approaches.

Large-scale studies for comparing machine learning and soft computing tools have focused on the classification domain [4]. On the other hand, very few extensive studies can be found in the regression domain. In [5], Nesrren et al. carried out a large-scale comparison of machine learning models for time series forecasting. The study includes techniques such as K-nearest neighbors (KNN), CART regression trees, multilayer perceptron networks, support vector machines, Gaussian processes, Bayesian neural networks, and radial basis functions. The research reveals significant differences between the methods studied and concludes that the best techniques for time series forecasting are multilayer perceptron and Gaussian regression when applied on the monthly M3 time series competition data (a thousand-time series) [6]. Moreover, in [7], an empirical comparison of regression analysis is carried out, as well as decision trees and artificial neural network (ANN) techniques for the prediction of electricity energy consumption. The conclusion was that the decision tree and the neural network models perform slightly better than regression analysis in the summer and winter phases, respectively. However, the differences between the three types of models are quite small in general, indicating that the three modeling techniques are comparable when predicting energy consumption.

In recent studies, the good omen of AI techniques is shown. For instance, Jurado et al. [8] propose an entropy-based feature selection process to select the most important past consumption values and it is combined with machine learning and soft computing techniques. Moreover, these hybrid methodologies are compared against a typical statistical technique (ARIMA), resulting in higher accuracies for FIR, random forest, and artificial neural networks (NNs). Thus, hybrid methodologies combine the strengths of different techniques to achieve higher accuracies in predictions. The results showed in this study also highlight the adaptability and scalability of these models to buildings with different profiles (location, usage, etc.).

In addition, in contrast to other approaches where offline modeling takes considerable computational time and resources, the models discussed in [8] appear to generate fast and reliable models, with low computational costs. These models can be embedded, for instance, in a second generation of smart meters where they could generate on-site forecasting of the consumption and/or production in the next hours or even trade the excess energy with other smart meters.

Considering the strengths of some AI methodologies reported in the literature (prediction of unexpected changes, predict energy consumptions and productions in different types of buildings, etc.) and the experience inside the consortium with such techniques, AI models are the approach followed for the *generation and consumption forecast*.

7.2.1.1 Random forest

Random forest (RF) is a set of CART, which was first put forward by Breiman [9]. In RF, the training sample set for a base classifier is constructed by using the Bagging algorithm [10]. In traditional CART, each inner node is a subset of the initial data set and the root node contains all the initial data. RF is a combination of tree predictors such that each tree depends on the values of a random vector sampled independently and with the same distribution for all trees in the forest. RFs for regression are formed by growing trees depending on a random vector such that the tree predictor takes on numerical values as opposed to class labels. The RF predictor is formed by taking the average over B of the trees. Figure 7.1 shows a scheme of the random forest.

Assuming the following basic notations:

- Let the number of training cases be N and the number of variables be P
- Input data point $\rightarrow v = (x_1, \ldots, x_P) \in \mathbb{R}^P$
- Output variable $\rightarrow$ c
- Let the number of trees be B

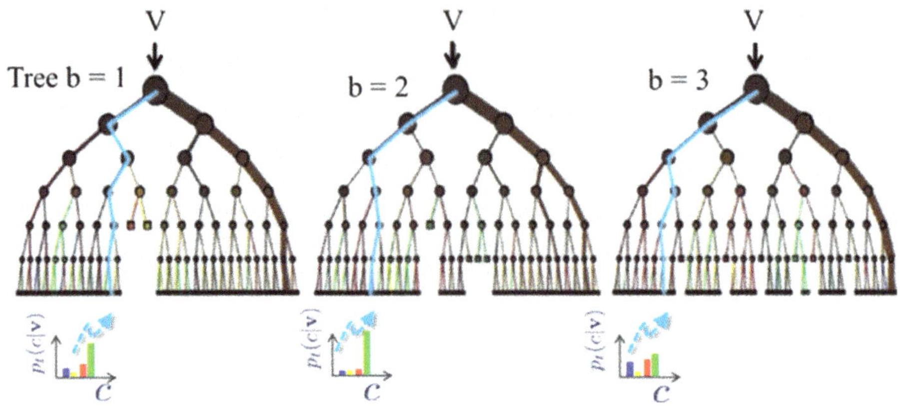

Figure 7.1 Random forest scheme containing three different trees.

The schematic RF algorithm is the following:

1. For $b = 1$ to B
    a. Draw a bootstrap sample Z^* of size N from the training data
    b. Grow a random forest tree T_b to the bootstrapped data, by recursively repeating the following steps for each terminal node of the tree, until the stopping criteria is reached:
        i. Select m variables at random from the P variables
        ii. Pick the best variable/split point among the m
        iii. Split the node into two daughter nodes
2. Output the ensemble of trees $\{T_b\}_1^B : p(c \mid v) = \frac{1}{B}\sum_1^B p_b(c \mid v)$

The size N of the bootstrap sample Z^* can go from a small size to the size of the whole data set. However, with large training data sets, assuming the same size can significantly affect computational cost. In addition, for big data problems such as the forecasting of consumption/production of all the buildings in a city, a good definition of the parameter N is mandatory. Moreover, there are different stopping criteria; two of the most commonly used are as follows: (1) until the minimum node size $n_{\min}$ is reached and (2) when a maximum tree depth is reached.

Although RF has been observed to overfit some data sets with noisy classification/regression tasks [11], it usually provides accurate results, generalizes well, and learns fast. In addition, it is suitable to handle missing data and provides a tree structured method for regression [12].

7.2.1.2 Artificial neural network

NNs are a very popular data mining and image processing tool. Their origin stems from the attempt to model the human thought process as an algorithm which can be efficiently run on a computer. Its origins date back to 1943, when neurophysiologist W. McCulloch and mathematician W. Pitts wrote a paper on how neurons might work [13], and they modeled a simple NN using electrical circuits. Some years later, in 1958, F. Rosenblatt created the perceptron, an algorithm for pattern recognition based on a two-layer learning computer network using simple addition and subtraction [14].

Many time series models are based on NN [15]. Despite the many desirable features of NNs, constructing a good network for a particular application is a nontrivial task. It involves choosing an appropriate architecture (the number of layers, the number of units in each layer, and the connections among units), selecting the transfer functions of the middle and output units, designing a training algorithm, choosing initial weights, and specifying the stopping rule.

It is widely accepted that a three-layer feed forward network with an identity transfer function in the output unit and logistic functions in the middle-layer units can approximate any continuous function arbitrarily well given sufficient amount of middle-layer units [16]. Thus, the network used in this research is a three-layer feed forward network (Figure 7.2). The inputs are connected to the output via a middle layer.

When working with univariate time series, the neurons of the input layer contain the present value and the relevant past values of the univariate time series, while the output is the value for the next time period, computed as described in Equation (7.1).

$$S(t+1) = f(s(t), \ldots, s(t-n)) \tag{7.1}$$

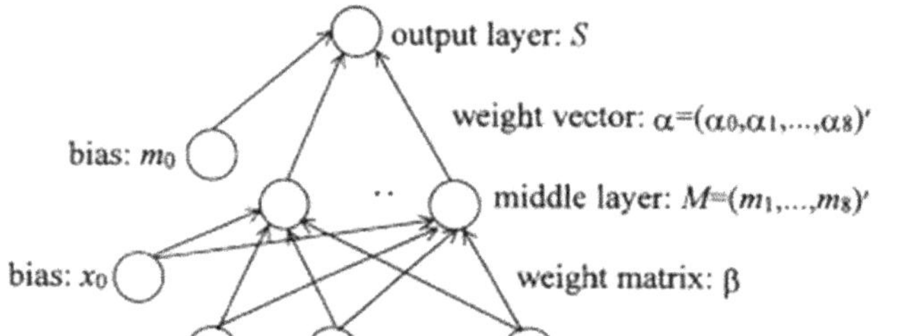

Figure 7.2 Representation scheme of a three-layer feed forward neural network.

where n is the number of past values of variable s and f is a nonlinear function approximated by a multilayer feed forward neural network (FNN) [17].

Recurrent neural networks are able to obtain very good prediction performance, since their architecture allows that the connections between units form a directed cycle, which allows it to exhibit dynamic temporal behavior. Unlike feed forward NNs, recurrent networks can use their internal memory to process arbitrary sequences of inputs. Therefore, recurrent NNs are very powerful, but they can be very complex and extremely slow compared to feed forward networks. As mentioned earlier, one of the main objectives of this research is to find powerful prediction methodologies with low computational costs, which could be embedded in a smart meter and generate on-site forecasting of the consumptions and/or productions. This is the reason why recurrent NNs have been rejected in this work and a cooperative approach has been chosen instead, i.e., FSP and a feed forward neural network that uses, as input variables, the most relevant past consumptions values.

7.2.1.3 Fuzzy inductive reasoning

The conceptualization of the fuzzy inductive reasoning (FIR) methodology arises from the general system problem solving (GSPS) approach proposed by Klir [18]. This methodology of modeling and simulation has the ability to describe systems that cannot be easily described by classical mathematics or statistics, i.e., systems for which the underlying physical laws are not well understood [19]. A FIR model is a qualitative nonparametric model based on fuzzy logic. Visual FIR is a tool based on the FIR methodology (runs under Matlab environment), which offers a new perspective to the modeling and simulation of complex systems. Visual FIR designs process blocks that allow the treatment of the model identification and prediction phases of FIR methodology in a compact, efficient, and user-friendly manner [20]. The FIR model consists of its structure (relevant variables or selected features) and a pattern rule base (a set of input/output relations or history behavior) that are defined as if-then rules. Once the best structure (mask) has been identified, it is used to obtain the pattern rule (called behavior matrix) from the fuzzified training data set. Each pattern rule is obtained by reading out the class values through the "holes" of the mask (the places where the mask has negative values) and place each class next to each other to compose the rule.

Once the behavior matrix and the mask are available, a prediction of future output states of the system can take place using the FIR inference engine, as described in Figure 7.3. This process is called qualitative simulation. The FIR

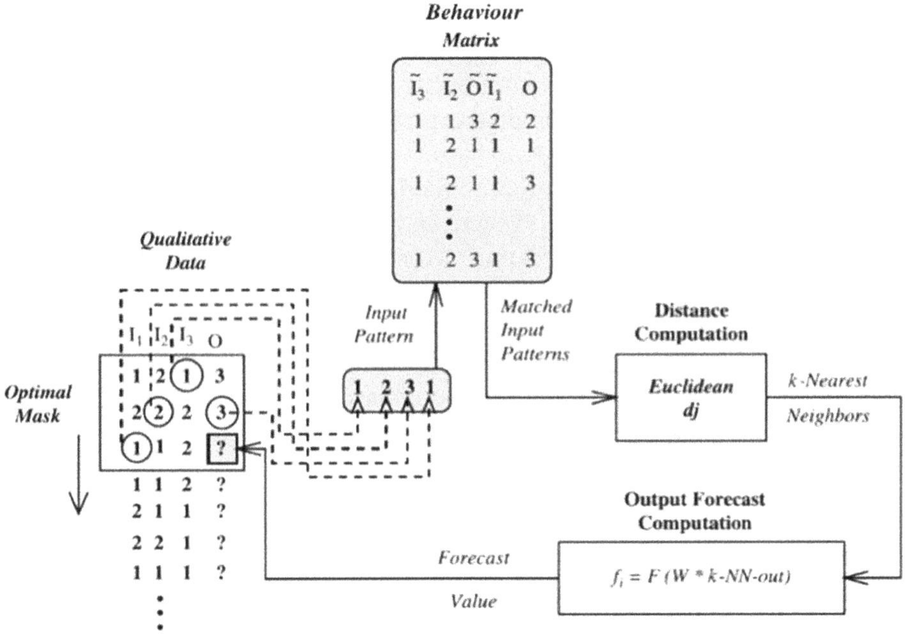

Figure 7.3 Qualitative simulation process diagram (with an example containing three inputs and one output).

inference engine is based on the KNN rule, commonly used in the pattern recognition field. The forecast of the output variable is obtained by means of the composition of the potential conclusion that results from firing the *k* rules whose antecedents have best matching with the actual state.

The mask is placed on top of the qualitative data matrix (fuzzified test set), in such a way that the output matches with the first element to be predicted. The values of the inputs are read out from the mask and the behavior matrix (pattern rule base) is used, as it is explained latter, to determine the future value of the output, which can then be copied back into the qualitative data matrix. The mask is then shifted further down one position to predict the next output value. This process is repeated until all the desired values have been forecast. The fuzzy forecasting process works as follows: The input pattern of the new input state is compared with those of all previous recordings of the same input state contained in the behavior matrix. For this purpose, a normalization function is computed for every element of the new input state and an Euclidean distance formula is used to select the KNN, the ones with smallest distance, that are used to forecast the new output state. The contribution of each neighbor to the estimation of the prediction of the new output state is a function of its proximity. This is expressed by giving a distance weight to each neighbor, as shown in Figure 7.3. The new output state values can be computed as a weighted sum of the output states of the previously observed five nearest neighbors.

The FIR methodology is, therefore, a modeling and simulation tool that is able to infer the model of the system under study very quickly and is a good option for real-time forecasting. Moreover, it is able to deal with missing data as has been already proved in a large number of applications [19]. On the other hand, some of its weaknesses are that as long as the depth and complexity increase, the computational cost increases too, and also the parameters to choose during the fuzzification phase (which can be mitigated using evolutionary algorithms to tune the parameters).

7.2.2 AI Generation and Consumption Forecast

7.2.2.1 Model generation

After an empirical analysis, we have finally selected RF for the first version of the prediction models. In contrast to most of the approaches previously explained, where offline modeling takes considerable computational time and resources, the technique used in the prediction algorithms, RF, appear to generate fast and reliable models, with low computational costs.

FIR is also a suitable technique with higher accuracies than RF but its development and implementation are more difficult than RF.

To perform a prediction, it is necessary to first generate a model. There are multiple strategies for the model generation, for instance to cluster similar installations and create a single model that would fit with all the installations within this cluster. The main advantage of this strategy is that only one model is created, and therefore, the computational cost and the storage capacity needed are very low. On the other hand, the accuracy may be lower than the approach, *one model one installation,* where a single model is created for each installation. In iURBAN, the model generated is unique for each installation. This would not be the best solution for a pilot with thousands of installations, because it would take high computational costs and increase storage capacity needs. However, since the iURBAN pilot has around 100 installations, the approach *one model one installation* is an affordable solution.

In Figure 7.4, the process to generate a prediction model and the online predictions is explained. In the left side of the figure, the *Model Generation* (1) is created/updated every week as default. The model is built-up taking all the historical past consumptions from the *Data Warehouse* (2) in the cloud.

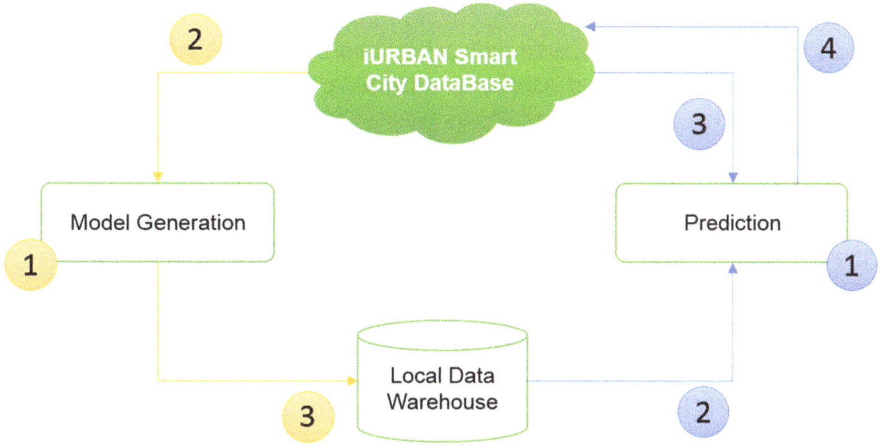

Figure 7.4 Prediction algorithm flow diagram.

After the model is created/updated, it is stored in the *Local Data Warehouse* (3) of the PC/server where it is executed.

On the other hand, in the right side of Figure 7.4, the *Prediction* (1) is performed every hour as default. When a prediction model is available in the *Local Data Warehouse,* it is loaded and used for the online prediction (2). The prediction is performed taking the last hours/days historical past consumptions from the *iURBAN SCDB* (3). After the prediction is performed, the results are sent to the *iURBAN SCDB* (4), where are available for any component in iURBAN.

7.2.2.2 Model and prediction configuration parameters

The granularity of the predictions is an important parameter because it defines the prediction intervals. Depending on the use case of the prediction data, granularity will be determined. For example, if the use case is to analyze potential blackouts, variability of renewables in the electricity grid, solar, and wind forecasting or real-time demand fluctuations, then the granularity level should be high because it requires a prediction at second level or even millisecond. This is not the case of iURBAN. On the other hand, if the use case is to create awareness about possible consumption/production at the end of the day, week, and/or month, at which hour is expected a peak, detect a day where higher consumptions or productions are expected, etc., then lower levels of granularity are needed. This is the case of iURBAN.

Highest granularity of data in Plovdiv and Rijeka is 15 min and 1 min, respectively. The smart algorithm modules are developed to accept any granularity level; however, based on the initial requirements of iURBAN, specifications, and LDSS/CDSS functionalities and graphical user interface (GUI), it has been decided that a good compromise of prediction interval is 60 min. The model is updated once per week.

7.2.2.3 Grids and levels

Prediction models run for master sensors of water, heating, gas, and electricity, as well as for global production of electricity and heating. Master sensors are those sensors measuring the total consumption or production in apartments, buildings, substations (heating), and combined heat and power (CHP) plants.

7.2.3 Development and Implementation

7.2.3.1 Code

The code was developed in Java. It consists of six different packages:

- *iurban.predictionengine.webapi* (contains one class):
  - *WebApiConsumer*. This class contains all the methods to interact with the SCDB; *HttpPost* and *HttpGet* to download historical consumption data and upload forecasting values.
- *iurban.predictionengine.models* (contains 5 classes):
  - *Attributes, Gap, Granularity, Outlier, Result, Stats, WeatherAnalytics, Sensor, Prediction, Installation, TimeSeries*. These classes contain the data structure to handle correctly the data obtained through the API.
- *iurban.predictionengine.utils* contains 15 classes. These classes contain common methods used in several processes of the model generation and prediction.
  - *ArrayUtils, Comparators, ErrorUtils, VectorUtils, WekaUtils, PropertiesUtils, etc.*
- *iurban.predictionengine.predictionmodel* that contains the two main classes for the prediction values generation:
  - *WekaTechnology*. This class contains the technology to design and generate the models in Weka.
  - *FirTechnology*. This class contains the technology to design and generate the models with FIR technique.
  - *PredictionOnTheFly*. Within this class, there are all the methods to perform the predictions with the model already generated.

- *iurban.predictionengine.prediction* is the package with the classes containing the main programs for the model generation and forecasting values generation.
  - *PredictionModelMain.* Class containing the main to create the models.
  - *PredictionOnTheFlyMain.* Class containing the main to generate the predictions.

7.2.3.2 Deployment

In order to deploy the AI forecasting models, it is necessary to first generate a *jar file* containing all the aforementioned classes, as well as the necessary libraries. The *jar* file can run in any PC or local/cloud server that has a Java Virtual Machine.

It has to be taken into account that the models generated are stored locally in the machine where the *jar* file is running.

Along with the *.jar* file, it is necessary to create a *config.properties* file containing the following information:

DirectoyNameModel: Path where the models generated are stored.

URL: Web API.

User: iURBAN user with credentials to use the data.

Password: Password of the user.

Aggregation: Desired aggregation for the prediction in minutes. For instance, in the example above, the models and predictions are performed with daily (aggregation of 1440 min, which is equal to 24 h) and hourly data (aggregation of 60 min, which is equal to 1 h).

```
DirectoryNameModel=C:\\SensorModels\\
URL=http://iurban.azurewebsites.net
User=username
Password=password
Aggregation=1440,60
WindowForecasting=31,72
PredicitonCategory =master,gasmaster,heatingmaster,watermaster
PredictionType=ProfileElectricityProduction,ProfileHeatEnergyProduction,ProfileHotWaterProduction
SleepTimeExecutionModel=1440
SleepTimeExecutionPredictionOnTheFly=60
InstallationsToStudy=all
InstallationsToDiscard=
Attributes=Consumption,WeekDay,TempOut,Hour,PreviousValue,PreviousWindowForecastingValue
ExternalWeatherInfo=yes
```

Figure 7.5 config.properties file example.

WindowForecasting: Horizon of the prediction. In the example above, the first value ("31") corresponds to the aggregation value "1440"; therefore, it is a daily prediction for the next 31 days. The second value ("72") corresponds to the aggregation value "60"; thus, it is an hourly prediction for the next 72 hours.

PredicitonCategory: ConsumptionCategories of sensors that will be predicted.

PredicitonTypes: Type of sensors that will be predicted.

SleepTimeExecutionModel: How often the models are updated in minutes.

SleepTimeExecutionPredictionOnTheFly: How often the prediction is performed in minutes.

InstallationsToStudy: Installations in the data warehouse to perform the prediction.

InstallationsToDiscard: Installations discarded to perform the prediction.

Attributes: Attributes used to create the models and perform the predictions.

ExternalWeatherInfo: Use of weather information to train the model and predicts.

The deployment of the AI forecasting models has been done in the cloud. The cloud service has been subcontracted to *Microsoft Azure.* The specifications of the instance are as follows:

- Windows server 2012 R2
- Number of CPU cores: 2
- RAM: 3.5 GB
- Local resource: 496.664 MB (496 GB)

7.3 Dynamic Tariff Comparison and Demand Response Simulation

7.3.1 Functionality

Dynamic tariff comparison and demand response simulation functionality are implemented as VPP functionality. The VPP is a back end calculation engine only. This means that there is no direct interaction between iURBAN tool users and the VPP. Users interact with the VPP via the CDSS GUI.

Based on historical consumption data for one or more city model prosumer objects, the VPP provides following functionality:

- Calculate energy costs for different dynamic tariffs[1] (in the following called *test tariffs*) against a tariff currently in place.
- Account for potential user behavior changes: Simulate how dynamic tariffs could influence consumption patterns of the selected city object(s)—e.g., reduced consumption during high price hours, increased consumption during low price hours (based on T3.3 work).

7.3.2 Stimulus/Response Sequence

The user interacts with the CDSS GUI to initiate tariff comparison and demand response simulation calculations. The user's simulation request is passed to the VPP via the SCDB-CDSS as shown in Figure 7.6.

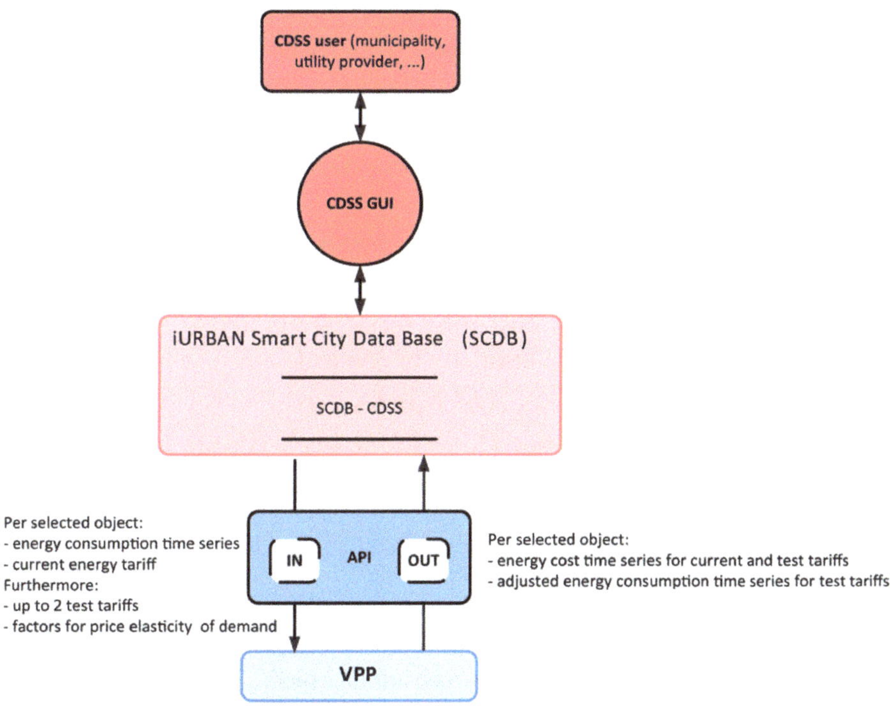

Figure 7.6 Data flow diagram: dynamic tariff comparison and demand response simulation.

[1]Note: In the context of iURBAN, the term "dynamic tariff" simply means that energy cost is a function of time. Dynamic tariffs are defined well in advance (e.g., adjusted only once every X months) and do not change in the short term. Other than dynamic tariffs, "(near-)real-time" tariffs change in the short term (e.g., announced 24 h in advance for the next day based on renewables generation forecast). "(Near-)real-time" tariffs are out of scope of iURBAN.

Per selected prosumer object *i*, the VPP expects the following input data:

- "As is" energy consumption time series (historical data): $cons\_as\ is_i(t)$
- "As is" energy tariff currently in place as a function of time: $tariff\_as\ is_i(t)$.

Optional input data per prosumer object *i*:

- Up to two alternative *test tariffs* selected for comparison $tariff\_test_{i,j}(t)$,
- Three floating point values representing low, medium, and high price elasticity of demand: $elast_i = (elast_{i,\text{low}},\ elast_{i,\text{medium}},\ elast_{i,\text{high}})$.

If energy consumption or tariff data are missing for any of the selected city objects, the VPP does not carry out a calculation and returns an error message.

If all required data are available, the VPP carries out calculations and provides the following results data per selected prosumer object *i*:

- "As is" energy consumption time series: $cons\_as\ is_i(t)$
- "As is" energy cost time series: $cost\_as\ is_i(t)$
- If *test tariff(s)* (up to 2) have been supplied, per test tariff:
    - 3 energy consumption time series according to low, medium, and high price elasticity of demand
        - $cons\_test\_low_i(t)$, $cons\_test\_medium_i(t)$, $cons\_test\_high_i(t)$
    - 4 energy consumption cost time series according to none, low, medium, and high price elasticity of demand
        - $cost\_test\_none_{i,j}(t)$, $cost\_test\_low_{i,j}(t)$, $cost\_test\_medium_{i,j}(t)$, $cost\_test\_high_{i,j}(t)$

7.3.3 User Workflow

1. **CDSS** user actions:
    a. The user opens the CDSS GUI tariffs analysis module.
    b. The user selects a commodity of interest (e.g., electricity or gas).
    c. The user selects one or more prosumer object, i.e., buildings or other city objects of interest.
    d. For each prosumer object selected, the user assigns
        i. The current tariff (note: The CDSS is to provide a tariff database so that a user can easily select tariffs previously defined).

ii. 0 to 2 alternative tariffs (*test tariffs*) (note: To speed up the process, the CDSS GUI probably provides functionality to assign the same tariff to all selected objects).
iii. Optional: Values for low, medium, and high elasticity of demand (note: The CDSS GUI provides default values per commodity).

e. The user saves the simulation request under a specific name.

2. User input is stored in the **SCDB-CDSS** and parsed to the VPP from the SCDB-CDSS.
3. **VPP** actions:
   a. A new calculation job is created and added to the job queue.
   b. The simulation job is carried out.
   c. Simulation job results parsed back to the SCDB-CDSS.
4. The simulation results are stored in the **SCDB-CDSS**.
5. **CDSS user** actions:
   a. The user analyses simulation results.
   b. The user can select to see aggregated results for a selected group of objects. Aggregation calculations required for this are carried out by the CDSS.

7.3.4 Calculation Methodology

7.3.4.1 Price elasticity background

Price elasticity of demand is a simple concept used in economics to estimate the change of demand of a certain good if its price changes.

It is defined as follows:

$$\text{Price elasticity} = \frac{\%\triangle\ \text{Quantity Demanded}}{\%\triangle\ \text{Price}}$$

University of Freiburg carried out research on how well this concept can be applied in the context of energy demand.

Suggested reasonable ranges for elasticity factors for electricity and gas:

	Residential Electricity	Commercial Electricity	Residential Natural Gas
Short-run elasticity	–0.24	–0.21	–0.12
Long-run elasticity	–0.32	–0.97	–0.36

Absolute values of long-run elasticity were found to be higher than short-run ones. This indicates that users seem to require some time to adjust to different prices.

Note 1: When applying elasticity factors, the total consumption changes:

$$total_{consumption(real_{tariff})} \neq total\_consumption(test\_tariff).$$

This is in agreement with real-world observations: Customer DR initiated by flexible tariffs has both load shifting and energy saving/"wasting" components:

- Load shifting: Equipment is used at a different time of the day, but in total, the same amount of energy is consumed.
- Energy saving: Consumption is reduced, e.g. by dimming light intensity during high-price periods.
- Energy "wasting": Consumption is increased, e.g., do not switch off the light in non-occupied rooms because electricity is cheap anyway.

7.3.4.2 Dynamic tariff comparison and demand response formula

If all required data are available, the VPP carries out calculations and provides the following results data per selected prosumer object *i*:

Demand response:

"As is" consumption

$$= cons\_as\ is_i(t)$$

Furthermore, if test tariffs (up to 2) have been provided consumption time series adjusted by price elasticity (low, medium, high) of demand:

$$cons\_test\_low_{i,j}(t) = \left(\left(elast_{\ i,low} \times \left(\frac{tariff\_as\_is_i(t)\ -\ tariff\_test_{i,j}(t)}{tariff\_test_{i,j}(t)}\right)\right)\right. \times cons\_as\ is_i(t)\Bigg) + cons\_as\ is_i(t)$$

$$cons\_test\_medium_{i,j}(t) = \left(\left(elast_{\ i,medium} \times \left(\frac{tariff\_as\_is_i(t)\ -\ tariff\_test_{i,j}(t)}{tariff\_test_{i,j}(t)}\right)\right)\right. \times cons\_as\ is_i(t)\Bigg) + cons\_as\ is_i(t)$$

$$cons\_test\_high_{i,j}(t) = \left(\left(elast_{\,i,high} \times \left(\frac{tariff\_as\_is_i(t) - tariff\_test_{i,j}(t)}{tariff\_test_{i,j}(t)}\right)\right) * cons\_as\ is_i(t)\right) + cons\_as\ is_i(t)$$

Dynamic tariff:

"As is" energy cost:

$$cost\_as\ is_i(t) = cons\_as\ is_i(t) \times tariff\_as\ is_i(t)$$

Furthermore, if *test tariffs* (up to 2) have been provided:

$$cost\_test\_none_{i,j}(t) = cons\_as\ is_i(t) \times tariff\_test_{i,j}(t)$$
$$cost\_test\_low_{i,j}(t) = cons\_test\_low_{i,j}(t) \times tariff\_test_{i,j}(t)$$
$$cost\_test\_medium_{i,j}(t) = cons\_test\_medium_{i,j}(t) \times tariff\_test_{i,j}(t)$$
$$cost\_test\_high_{i,j}(t) = cons\_test\_high_{i,j}(t) \times tariff\_test_{i,j}(t)$$

7.3.5 Assumptions and Limitations

- In contrast to other VPP calculations, tariff comparison and DR simulation calculations are carried out for historical data only (considering forecast data as well would not provide extra value).
- The user may not define the time span for tariff analysis: Per default, up to one year of historical data is analyzed.
- Tariff analysis and DR calculations are to be carried out on prosumer object level only. This means that no aggregation to a higher level or re-calculation of VPP network status based on changed consumption patterns of city objects is required.
- The VPP shall provide some kind of job batch queue mechanism to allow a user to schedule several tariff comparison and demand response simulation runs. This implies that different tariff comparison and demand response jobs are carried out independently from each other. If a user requests to store results of more than one tariff comparison and DR simulation at a time, this needs to be implemented as functionality of the system connecting to the VPP (SCDB-CDSS).
- Tariff comparison and DR simulation consider energy consumption only; energy generation is not considered.[2]

[2]One reason for this decision is that feed-in tariff structures are comparatively complex in iURBAN demo cities. The project consortium agreed to focus on energy consumption only.

- The dynamic tariff data model only needs to support tariffs of a format which can be expressed as function of time (=time series). Aspects such as standing charges and capacity charges are not supported.

7.4 Conclusions

This chapter has presented the two main smart algorithms within iURBAN. Prediction of consumption and production of energy were demanded by households as a must feature, receiving good feedback about its representation through the graphical user interface.

Variable tariff and demand response simulation capabilities have been describing initial assumptions as well as a use case showing the expected iteration of end users with VPP (the tool which integrates the algorithms) through the CDSS interface.

References

[1] Khamis, M. F. I., Baharudin, Z., and Hamid, N. H. (2011). Electricity forecasting for small scale power system using artificial neural network. *Power Engineering and Optimization Conference (PEOCO) 5th International*, pp. 54–59.

[2] Marvuglia, A., and Messineo, A. (2012). Using recurrent artificial neural networks to forecast household electricity consumption. *Energy Procedia*, 14, 45–55.

[3] Tranchita, C., and Torres, A. (2004). Soft computing techniques for short term load forecasting. *Power Syst. Conf. Expo.*, 1, 497–502.

[4] Caruana, R., and Niculescu-Mizil, A. (2006). An empirical comparison of supervised learning algorithms. *Proceedings of the 23rd International Conference on Machine Learning (ICML2006)*, pp. 161–168.

[5] Nesreen, A. K., Amir, A. F., Neamat, G., and Hisha, E. H. (2010). An empirical comparison of machine learning models for time series forecasting. *Econometric Rev.,* 29, 594–621.

[6] http://www.forecasters.org/data/m3comp/m3comp.htm

[7] Tso, G. K. F., and Yau, K. K. W. (2007). Predicting electricity energy consumption: A comparison of regression analysis, decision tree and neural networks. *Energy*, 32, 1761–1768.

[8] Jurado, S., Nebot, A., and Mugica F. (2015). Hybrid methodologies for electricity load forecasting: Entropy-based feature selection with machine learning and soft computing techniques. *Energy*, 86, 276–291.

[9] Breiman, L. (2001). Random forests. *Machine Learning*, 45 (1), 5–32.
[10] Breiman, L. (1996). Bagging predictors. *Machine Learning*, 24 (2), 123–140.
[11] Kleinberg, E. (1996). An overtraining-resistant stochastic modeling method for pattern recognition. *Annals of Statistics* 24 (6), 2319–2349.
[12] Li, Y., Wang, S., and Ding, X. (2010). Person-independent head pose estimation based on random forest regression. *17th IEEE International Conference on Image Processing (ICIP)*, pp. 1521–1524.
[13] McCulloch, W., and Pitts, W. (1943). A logical calculus of ideas immanent in nervous activity. *Bullet. Mathematic. Biophy.*, 5(4), 115–133.
[14] Rosenblatt, F. (1958). The perceptron: A probalistic model for information storage and organization in the brain. *Psychol. Rev.*, 65(6), 386–408.
[15] Alon, I., Qi, M. and Sadowski, R. J. (2001). Forecasting aggregate retail sales: A comparison of artificial neural networks and traditional methods. *J. Retail. Consum. Serv.*, 8, 147–156.
[16] White, H. (1990). Connectionist nonparametric regression: multilayer feed forward networks can learn arbitrary mappings. *Neural Networks*, 3, 535–549.
[17] Chow, T. W. S., and Cho, S. Y. (2007). Neural Networks and Computing: Learning Algorithms and Applications. 7, 14–15.
[18] Klir, J., and Elias, D. (2002). *Architecture of systems problem solving*, 2nd. Edn. Plenum Press, New York.
[19] Nebot, A., Mugica, F., Cellier, F., and Vallverdú, M. (2003). Modeling and simulation of the central nervous system control with generic fuzzy models. *Trans. Society Model. Simulat.*, 79(11), 648–669.
[20] Escobet, A., Nebot, A., and Cellier, F. E. (2008). Visual-FIR: A tool for model identification and prediction of dynamical complex systems. *Simulat. Modell. Practice Theory*, 16, 76–92.

8

Solar Thermal Production of Domestic Hot Water in Public Buildings

Energy Agency of Plovdiv

Energy Agency of Plovdiv, Plovdiv, Bulgaria

Abstract

This current chapter addresses the energy production situation in the pilot city of Plovdiv as investigated under the iURBAN project co-funded under the FP7 call. The focus of the article is the case study of one of seven public buildings with solar thermal installations realised with national and municipal funding. The pilot buildings are socially significant facilities – public kindergartens, located per urban residential area. Their study draws conclusions over the effectiveness and usefulness of such installations in public buildings and builds upon their role as prosumers in the urban energy balance.

Keywords: Energy production, Prosumers, Domestic hot water, Energy efficiency, Renewable energy resources, CO_2 reduction.

8.1 Introduction

The current chapter addresses the energy production situation in the pilot city of Plovdiv as investigated by the iURBAN project. The focus is a case study of one of seven public buildings with solar thermal installations realized through national and municipal funding. The pilot buildings are socially significant—i.e., public kindergartens, distributed evenly in the pilot city residential areas. The study draws conclusions on the effectiveness and usefulness of solar monitoring installations in public buildings and builds upon their role in detecting prosuming and feeding the information to the urban energy balance.

8.1.1 The Pilot

The city of Plovdiv is the second largest city in Bulgaria, approximately 152 km southeast of the Bulgarian capital of Sofia. The population is over 376,000, and there are approximately 300,000 visitors per year, including 80,000 foreign tourists.

Within the iURBAN project, the pilot city of Plovdiv built energy monitoring and management facilities for both energy consumption and production in 30 public and residential buildings. In the research and development framework of the project, it applied novel ICT and cloud-based services for promoting energy efficiency and renewable energy sources utilization on its territory.

Even though the energy management and monitoring systems in Plovdiv are not new to the residential buildings, only a few public buildings operate such. The mission of the iURBAN project was to expand their use and bring new perspective on the building and urban energy balance. Its integrated and validated ICT energy management systems (Chapter 4) gathered data useful for the local energy planning and development.

Monitoring of energy production was never realized in the city of Plovdiv up till the implementation of the iURBAN project. Putting a new, prosumer perspective on the realization of energy efficiency in the public sector, the seven public kindergartens were chosen—they would produce domestic hot water (DHW) for their own use and thus indirectly reduce the need for centralized hot water provision.

8.2 Public Solar Prosumers Background

8.2.1 Background

Traditionally, solar thermal installations in the city of Plovdiv are realized in the residential sector for single-family houses. Still, their usefulness and effectiveness outreach to the public users.

The seven pilot prosumers were renovated in 2013 with national and municipal funding, i.e., energy efficiency measures were realized—new external wall insolation, change of windows, change of lighting, and refurbishment of heating installation, along with the introduction of solar thermal installations for DHW.

8.2.2 How Is the Energy Management and Monitoring Architecture Established?

In late 2014, an energy consumption and production monitoring and management installation was introduced by the iURBAN project.

The deployed solar prosumer smart metering system is independent and parallel to the utility energy consumption monitoring system. Its architecture is realized so that each prosumer kindergarten has its own smart solar meter with communication facilities to transmit data to a centralized data collection center which forwards the data to the iURBAN Cloud. The equipment includes communication modules and a *blind* PC to manage reading, writing, and transmitting data on site. It is possible to update PC software from only reading and transmission to managing the entire heating system. The *blind* PC then sends the data to a local centralized data collection center where the data are kept and forwarded to the iURBAN cloud.

This type of system reduces the transmission failure risk as the solar meter, the *blind* PC and the buffering server back up data before sending it to the iURBAN cloud. In addition, data granularity can be adjusted to any transmission pattern, which can be defined by both the data receiver and the data sender.

The parallel prosumer smart metering system is connected to the iURBAN data cloud, and though consumption and production data are sent from two different sources, this does not affect the data visualization on the iURBAN platform.

In the future, the prosumer system can be expanded beyond by registering new solar prosumers, so that it forms a regional system for monitoring of the heat production, thus to become municipal prosumer data center, relevant to the local authorities, public end users, and community.

8.3 Case Study of a Prosuming Kindergarten

8.3.1 Introduction

The kindergarten studied is a public kindergarten, with solar panels installed through national and municipal funding. The iURBAN system was built additionally to capture electricity, heating, and DHW consumption from the local utility and production of domestic hot water by solar thermal installation.

The solar system in the kindergarten has 12 solar vacuum pipe collectors, with a total number of 260 pipes in a system of mixed type. All of them are propylene glycol collectors, which influences the metering characteristics of the smart meter for the solar installation.

The kindergarten has two water tanks of 1 000l each and 2 serpentines per tank—one for propylene glycol and second one for the local utility heat supply. Additional heating of the water inside the tanks could be done through two electric heaters per tank, i.e., four electrical heaters in total.

8.3.2 What We're Interested in and How Data Can Tell It?

The current analysis is pointed toward the effectiveness of introducing solar thermal installations in a public building and the cost benefit of integrating monitoring system to it. We will be also looking at the relations between the usage of centralized DHW and own production of DHW and how this affect the general energy balance of the building. Finally, we will have a look at the overall financial benefits of the case and if they have implications in regard to their role as energy and finance saving measure.

For the purposes of these analyses, we will be using data from the baseline year 2012 and the monitoring year 2015. Data are acquired through distant metering in 2012 and real-time metering in 2015 through the iURBAN platform service. The data sets used for 2012 are electricity and heat energy consumption only. The data sets used for 2015 are electricity consumption, heat consumption divided into two subsets—heating energy and DHW energy, and solar DHW production. Only one data gap in March and April 2015 was identified—data were filled through normal distribution based on the working days. Data in January split into heating and DHW was not present so the distribution was made based on the closest month proportion; this approach was chosen to the annual average one due to seasonal concerns.

Weather data for the two periods are acquired through online service and then used for normalization of the heat data.

The evaluation is based on a comparison between 2012 where no energy efficiency and renewable energy sources measures were realized and 2015 where all measures were fully realized and operational.

8.3.3 What the Results Tell Us for Baseline and Post-retrofit Periods?

8.3.3.1 What was happening when no energy efficiency measure was implemented back in 2012?

The total annual energy consumption in 2012 was 472 MWh out of which 48 MWh (10%) was dedicated to electricity consumption and 425 MWh (90%) to heat consumption.

Figure 8.1 promptly shows the seasonal character of the heat energy use—high in the winter season and low in the summer. The reason for the high winter values is the lack of energy efficiency measures implemented and poor building condition.

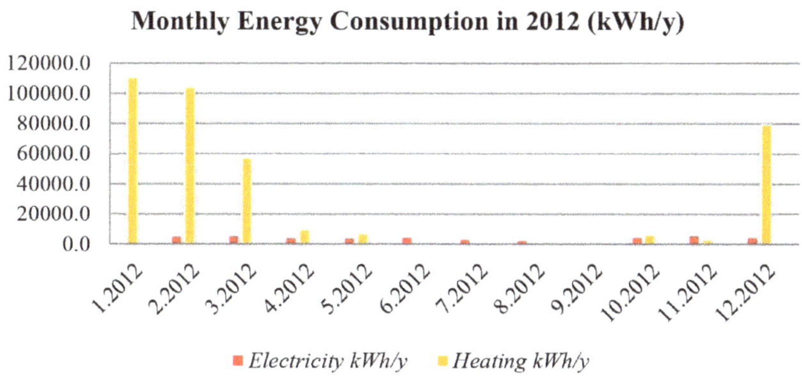

Figure 8.1 Monthly energy consumption in 2012 (kWh/y).

The correlation between outside temperature and the heat consumed in the kindergarten is –0.95, which corresponds to the prediction that with the rise of the outside temperature, the heat demand is reduced.

Figure 8.2 shows the share by energy type of the consumed energy. The share of electricity is 10% and that of the heat energy is 90%. The disproportion is due to the peculiarity of this type of public buildings—kindergartens spend most of their electricity energy for lighting and the kitchen needs, and the heat energy is used for keeping the indoor comfort up to the national requirements and domestic hot water needs in the kitchen.

In 2012, the electricity use in the kindergarten has produced 39.17 t CO_2 (24%) and the heat energy use 123.14 t CO_2 (76%).

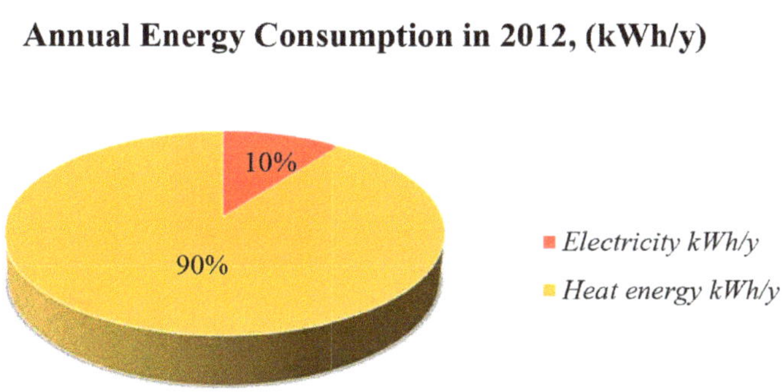

Figure 8.2 Annual energy consumption share by energy carrier in 2012 (kWh/y).

8.3.3.2 What happened when the building was deeply renovated and RES was introduced in 2015?

In summer 2013, energy renovation took place—insulation and new windows, and solar thermal installation for DHW was introduced. In autumn 2014, the iURBAN energy management system was introduced for energy consumption and production alike. Thus, the impact of energy efficiency measures and renewable energy sources introduction is visible throughout 2015.

The total annual energy consumption in 2015 is 282 MWh distributed as follows in Figure 8.3—electricity (21%), heat energy for heating (58%), heat energy for DHW from centralized heating supply (16%), and DHW from solar thermal production (5%).

The annual energy consumption and distribution in 2015 are shown in Figure 8.4. Winter months are high in heating demand (January, February,

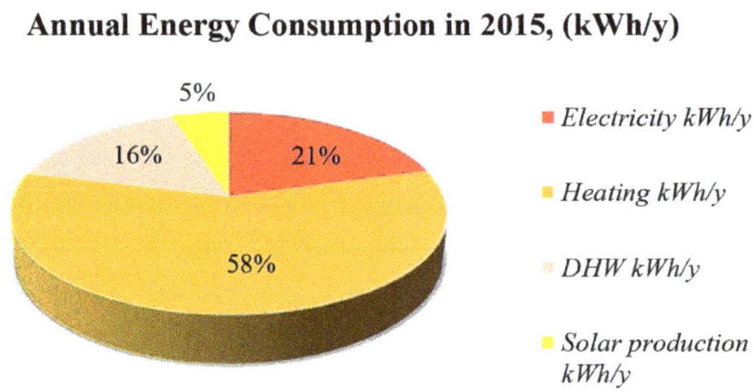

Figure 8.3 Annual energy consumption share by energy carrier in 2015 (kWh/y).

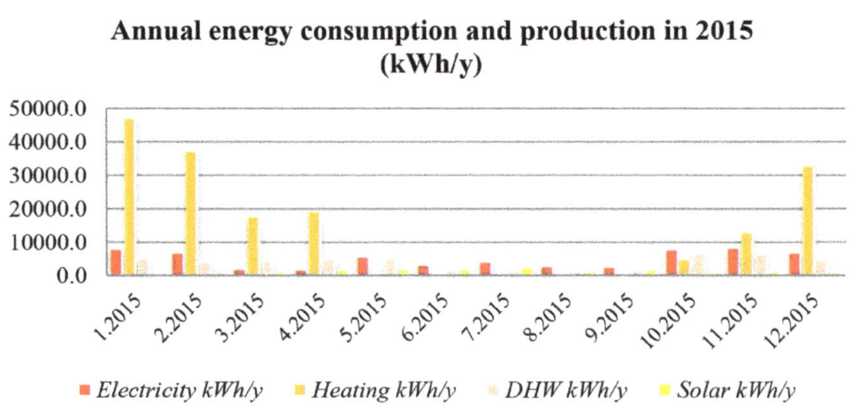

Figure 8.4 Monthly energy consumption in 2015 (kWh/y).

March, April, October, November, December). The correlation between outside temperature and the overall heat consumed in the kindergarten is –0.94, which is slightly lower than 2012, but again corresponds to the notion that the outside temperature and the heat demand are inversely proportional; looking at the heating demand only the correlation is –0.94 and at the DHW is –0.63.

On the other hand, DHW demand is constant throughout the year, but is also dependant on the number of children. In winter, the DHW consumption from centralized heat supply is 85% of the total annual DHW consumption and is due to the high number of children attending and lower capabilities of the solar installation to produce. In summer, the consumption is 58% of the total annual and corresponds to lower number of children attending and the increased capacity of the solar installation to produce.

The DHW produced by the solar thermal installation accounts for 23% of the total DHW consumption. Its RES nature and the technology used define low seasonality of the production—data show that production in the winter season is 44% and in summer season is 56% of the total annual production. Regarding the reduction in the DHW by centralized supply, the solar installation accounts for 15% of the DHW in winter and for 42% in summer. The correlation between the outside temperature and the solar production is 0.76 which shows direct positive effect of the warm weather on the production performance. In 2015, the correlation between DHW usage and solar production is –0.43.

In 2015, the CO_2 emissions by the kindergarten show significant reduction. The electricity use produced 47.81 t CO_2; the heating use, 47.57 t CO_2; and the domestic hot water use, 13.29 t CO_2 including the RES savings. The solar installation produced energy that saved 3.99 t CO_2.

8.3.3.3 So how did EE and RES measures bring change in the kindergarten energy balance?

The weather data are acquired through the local meteo online service performance. Based on the values, outside temperatures are normalized and so is the heating energy consumed. The heat consumption is reduced due to the introduction of the energy efficiency measures, including iURBAN energy monitoring, with 337 kWh/m^2 or 68%. Normalized data are presented in Table 8.1:

Normalization for the electrical energy use is not needed as the kindergarten does not heat or cool on electricity. Still, the electricity use has risen from 49.14 kWh/m^2 to 59.97 kWh/m^2 or 22% (Table 8.2).

Table 8.1 Normalized heating energy

	Heating Area	Heating Energy Consumed		HDD for This	Calibrated Heating Energy
Year	m^2	kWh/y	HDD	Climate Zone	kWh/m^2
2012	973.28	424,630	2552.5	2241.8	496.75
2015	973.28	164, 045	2093.8	2241.8	157.42

Table 8.2 Normalized electrical energy

	Heating Area	Electricity Consumed	Calibrated Electrical Energy
Year	m^2	kWh/y	kWh/m^2
2012	973.28	47,826	49.14
2015	973.28	58,371	59.97

Thus, the energy intensity per person per square meter of the kindergarten is 1.82 $kWh/m^2/p$ in 2012 and 1.09 $kWh/m^2/p$ in 2015, i.e., 40% reduction.

Table 8.3 shows that the overall CO_2 emissions produced in 2012 are 162.31 t CO_2 and in 2015, 108.67 t CO_2, i.e., 0.2t CO_2 per kid. This corresponds to overall reduction of 33% of the CO_2 emissions—there is 22% increase in the emission from electricity use and 51% reduction in the emissions from heat energy use. The solar energy in 2015 has 3.99 t CO_2 not realized which is 7% of the heat energy.

8.3.3.4 What is the overall impact of becoming a prosumer?

The solar installation and its monitoring infrastructure contribute significantly to the reduction in energy intensity and overall CO_2 savings in the kindergarten.

The particular solar installation in the studied kindergarten has produced 13.77 MWh energy for DHW in 2015 only. This corresponds to 3.99 t CO_2 saved and has brought a financial benefit of 607.54 EUR in 2015. The investment for establishment of the monitoring system is 1 246 EUR and will

Table 8.3 CO_2 emissions overview

	Electricity	Heat Energy: Heating	Heat Energy: DHW	Total t CO_2 Produced
Year	t CO_2	t CO_2		t CO_2
2012	39.17	123.14		162.31
2015	47.81	60.86		108.67
		47.57	13.29	

require 10% upon-demand technical support cost per year in the future. Its reimbursement will take 2, 4 years with the technical support.

8.3.4 Discussion

The current study brought up real-time data acquired through iURBAN energy management and monitoring platform to show the benefits of introducing energy efficiency measures and utilization of renewable energy resources. The iURBAN architecture and data gathering provide reliable, qualitative, and quantitative approach toward evaluating the introduction of energy measures.

The introduction of iURBAN provides direct observation over the technical status of the installations and the reduction in the energy consumption. The presence of the iURBAN software is valuable for regular check of the technical status of the solar installation and its functioning, i.e., additional setup of the system could bring up to 50% improvement in the solar energy production. In addition, iURBAN could be used to detect energy leaks—lighting left switched on, heating being on during weekends, etc., and thus to reduce energy costs.

Moreover, there could be direct evaluation of the impact—in the case of the studied kindergarten, it is 51% heat energy reduction, 22% increase in the electricity use, and general energy reduction of 33%. The introduction of the solar thermal installations was estimated to account for 23% of the DHW supply and the CO_2 emissions.

Relating to the cost-benefit of the iURBAN architecture and service, in this case, a reimbursement period of 2.4 years was calculated, but this high reimbursement is due to the low readiness of the solar installation system and additional technical setup made. In the future, other facilities could achieve even lower reimbursement period, which ***strongly justifies*** the introduction of real-time energy monitoring and management systems.

8.4 Conclusion

The introduction of the iURBAN architecture and services in the city of Plovdiv proved valuable novel experience for the public users. It proved significant energy reduction results and facilitated for further improvement of their energy status. In the studied kindergarten, iURBAN uncovered energy reduction of 40% and CO_2 reduction of 33%, and 6% contribution by RES utilization to the energy balance and 23% to the CO_2 emission reduction.

It also spurred new setup of the kindergarten energy systems and planning for new energy efficiency measures.

In the future, based on the iURBAN platform feedback, different prosumer profiles will be identified, thus facilitating optimization through prediction models and dynamic and demand response tariffs. The use of the local decision support system by the prosumers engages them, thus bringing them to aim higher at introducing more energy efficiency measures and achieving greater energy savings.

9

Business Models

Stefan Reichert and Jens Strüker

University of Freiburg, Freiburg im Breisgau, Germany

Abstract

In this chapter, we describe the potential impacts of the use-cases on the business activities of the involved market actors as well as the implications from testing the smartDSS functionalities in the context of the Bulgarian and Croatian energy market. It can be concluded that the use of smartDSS assists the daily business of utilities, municipalities, and probably also energy managers and creates new opportunities for their business models in the sense of providing the ground on which new services can be built. Regulatory barriers in the energy markets of Croatia and Bulgaria remain a significant constraint for exploiting the full potential of the energy management system.

Keywords: Business Models, Energy Management system, Regulatory Framework, Demand-Side Management.

9.1 Introduction

The increasing digitization in most aspects of today's society is one of the fundamental changes for the daily lives of all involved individuals. The so-called Internet of things (IoT) allows people and devices to connect with each other at any given time independent from their current location. This opens up a wide array of opportunities of how and when we can access information. These developments are mainly targeted to increase customers' comfort, flexibility, and the potential to save money, as transaction and information costs are reduced notably. However, in some other cases, the increasing digitalization is becoming also a necessity. This becomes evident in energy sectors that are aiming to reach a high share of renewable energy. Many countries around the world have set ambitious targets to reduce greenhouse gas emissions and subsequently aim to replace fossil energy sources with renewable energies, such as wind and solar power. Besides the reduction in greenhouse gases, the transition of the energy sector is oftentimes linked to broader sustainability goals, such as supporting electric mobility or raising energy efficiency. By extending energy generation from renewable sources, countries can replace existing non-renewable energy sources, such as coal and nuclear power. For instance, Germany strives for a twofold change consisting of nuclear phase-out and a concurrent abolishment of most fossil fuels [1]. Local generation and the inclusion of new actors such as prosumers are key to achieve the energy efficiency targets for the next years.

However, rising shares of volatile renewable energies are also posing many challenges for the energy supply. A successful transformation of the energy sector is often seen as inseparably linked with a smarter grid, as the old paradigm that supply follows demand can no longer hold. Information systems (IS) may constitute a key element of a smarter grid, as they provide the tools for more accurate measurements and predictions. The installation of smart meters in buildings is one of the main parts of required infrastructure. Up to now, individual countries have made very different experience regarding the installation of such smart meters and the development of a smart grid. In order to provide these benefits and meet EU targets, the smart grid must be able to seamlessly integrate various existing and/or new technologies—meters, sensors, data processing systems, etc.—with the physical infrastructure required to generate, transmit, and distribute electric power [2].

Utilities and city authorities have long been using network systems such as SCADA to optimize resources and monitor assets to carry out preventative maintenance. However, the rich data sets generated and stored in these "silo" systems are found in a variety of formats and are not easily accessed by third parties, thus preventing the optimal management, control, and efficiency of many city services (i.e., utilities, security, health, transportation, street lighting, and local government administration). Therefore, new systems are required that are fully functional within a smart grid and allow different market actors to access required data without violating the privacy of connected consumers and prosumers.

9.2 Benefit Framework for the Operation of an Energy Management Platform

9.2.1 Evaluation Framework

As with all information systems and technologies, the benefits of the implementation of an energy management system, such as the proposed smartDSS, need to be determined ex ante. We use a framework consisting of several benefit types for the identification of all potential business uses.

There are a number of existing methodologies for the quantification of benefits connected with the implementation of new technologies, such as decision support systems. For the smartDSS, there are a number of potential operators, which can yield different individual benefits. One common methodology suggests following a path similar to that applied to most other technological innovations [3]. This approach proposes the adjustment and application of well-established methods from the fields of capital budgeting and performance measurement. Mostly this method above addresses the problem of how to achieve overall quantification with the aim of establishing a basis for decisions regarding the initiation or postponement of an investment in a new technology (or, for that matter, decisions in respect of any investment accompanied by cost–benefit consequences that are insufficiently understood).

While some studies indirectly address ex ante benefit evaluation (namely field studies and live tests that require an actual implementation of the new technology), the remainder focus on only one or two of the three main aspects of benefit evaluation: classification frameworks, such as [4] help in identifying potential benefits of new applications without addressing quantification issues. Forecast models, such as the one presented by [5], address the forecast problem of investments, which is, on predicting the extent to which the number of processes and activities and/or resource consumption change, while largely neglecting the financial side. Finally, assessment models, such as [6] focus on the assessment problem, that is, on attaching a monetary value to multiple process improvements, while using "expert estimations" to bypass the forecast problem.

Based on the work from [7], the mentioned results can be adjusted in an ex ante evaluation framework that splits up potential benefits from the implementation of a decision support system into 3 benefit types:

Information: The measurement, collection, and visualization of production and consumption levels. As the smartDSS collects and displays detailed near real-time consumption and production data, it allows to create an overview of the current status of the grid and local areas. These information benefits do not require modified structures and processes. They might, however, include additional data gathering, which is not economically feasible without a smartDSS.

Optimization: Active balancing of production and consumption levels. As the smartDSS comprises both a centralized (CDSS) and local component (LDSS), it also allows to activate additional capacities, e.g., through demand-side management. This can happen both through immediate action (direct load control) and through long-term based changes in the pricing structure (e.g., the creation of new real-time tariffs). Equivalently, potential optimization benefits can comprise the optimal activation or deactivation of additional production and storage units.

Transformation: Re-engineering of existing business processes and investment decisions. Based on information and analytics functionalities of the smartDSS, decisions about investments into new infrastructure can be made, e.g., new solar panels in areas with high solar power potential or about the enhancement and reinforcement of the grid. Detailed information about consumption and production at all times allows an optimal market behavior and restructuring of the supply chain management.

While these three effects are helpful for identifying and categorizing benefits of the implementation of a decision support system, they offer only limited guidance for selecting and applying concrete instruments for an ex ante benefit quantification of a smartDSS operation. Therefore, these three benefit types can be further split up to help in identifying the related business potential in a case-specific setting and that are linked to specific quantification, i.e., forecast and assessment instruments.

Direct Benefits: Direct benefits of the operation of a smartDSS are understood to be effects that are immediately positive results of introducing the technology.

Indirect Benefits: Indirect benefits result from changes in decisions or systems that are enabled by a purposeful distribution and utilization of the data. They are therefore delayed in time and may be realized at different locations than where the data were collected.

As per definition, there exist no direct transformational benefits, as long-term changes within the organization or new investment decisions are only realized indirectly from the utilization of the data. These direct and indirect benefits can be further split up into operational and managerial benefits:

Operational Benefits: Operational benefits are created through improved work processes on a day-to-day basis. These benefits are therefore not based on a restructuring or implementation of new work processes but on a refinement of existing ones.

Managerial Benefits: Managerial benefits denote effects stemming from enhanced management support with integrated, aggregated, energy data which, in most cases, has been further refined and are targeted toward long-term positive results of the business processes. Managerial benefits require a data pool that is integrated and aggregated, and that stores historical data. Managerial benefits are by definition always indirect since they are based on data usage and are not realized in the short run (Figure 9.1).

9.2.2 Assessment of Benefits for Energy Providers

By sub-categorizing information, optimization, and transformation benefits further into direct/indirect and operational/managerial benefits, we derive in total eight benefit types for the operator of a Smart City Energy Management Platform, as visualized in Figure 9.2. As an operator, we thereby assume an energy provider or a utility that manages the delivery of energy to the end consumer.

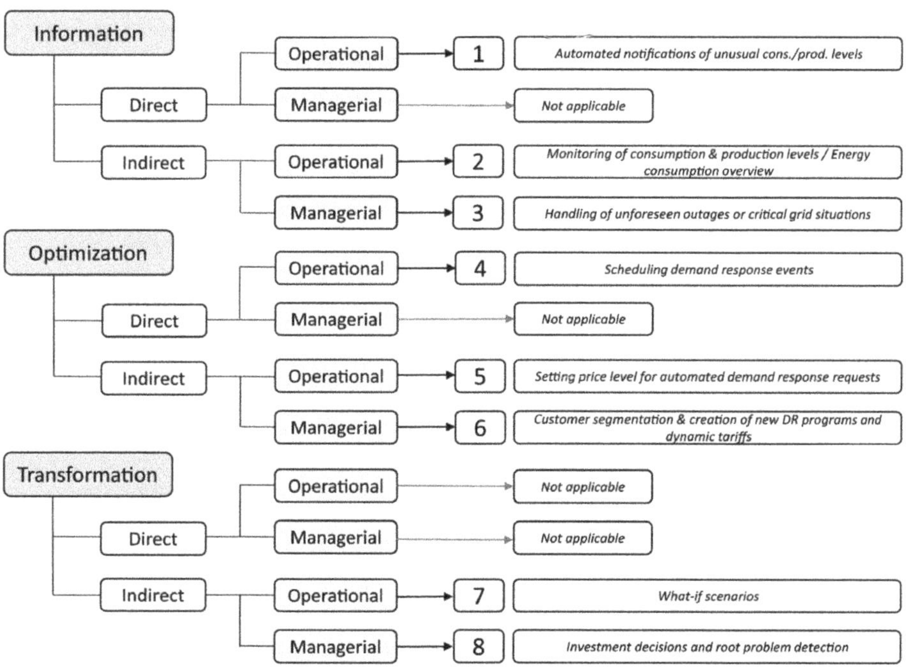

Figure 9.1 Segmentation of eight benefit types connected with the implementation of a Smart City Energy Management Platform.

Type 1—*Automated notifications of unusual consumption and production levels*: This benefit type can be directly generated from displaying information within the smartDSS and setting thresholds for which the system automatically gives a signal when it is reached. Unforeseen fluctuations in the energy production levels of solar and wind power installations can seriously jeopardize grid stability or cause significant costs to the energy provider by the activation of expensive balancing energy. Simultaneously, aggregated consumption levels can change unexpectedly, oftentimes in response to weather conditions as well. Automated notifications thereby help to anticipate critical grid situations. They lay the basis for further actions, e.g., optimization through sending out demand response requests.

Type 2—*Monitoring of consumption and production levels/energy consumption overview*: The smart meter data captures continuous consumption data, which enables the identification of inefficient consumption behaviors, especially for public buildings, since energy consumers here have little incentives to use energy efficiently, as they do not have to pay for the costs. For instance, public buildings may be kept heated on weekends, even though they are not used. The analysis of smart meter data helps to detect those inefficiencies in consumption patterns and significantly reduce energy costs. In contrast to benefit type 1, this benefit type is generated indirectly, through the manual analysis of data.

Type 3—*Handling of unforeseen outages or critical grid situations*: Based on the monitoring of production and consumption data, as well as their graphical display within the energy management system, specific bottlenecks within the grid can be detected. This allows to create strategies of how to handle situations, in which certain components of the smart grid become unavailable, e.g., during the replacement or failure of existing battery storages. Since the handling of unforeseen outages can cause very high costs, in the worst case through the blackout of the whole system, the required strategies can only be created on a managerial basis, through the careful evaluation of all existing data.

Type 4—*Scheduling demand response requests*: An active demand-side management can generally reduce the need to call expensive backup capacity in times of peak demand. For instance, some older power plants are no longer used on regular basis, but are infrequently activated to cover the peak demand on some days. This operation is not efficient both economically and environmentally. An active demand-side management as alternative to the activation of costly backup capacity can result in considerable savings for energy providers [8].

Type 5—*Setting price levels for automated demand response requests*: Based on the direct benefits from scheduling demand response events, the results from previous requests and customers' reactions can be analyzed for the adjustment of incentive schemes that are required for the participation of consumers.

Type 6—*Customer segmentation and creation of new demand response programs and dynamic tariffs*: Based on the benefit types 4 and 5, conclusions on a managerial basis can be taken regarding the segmentation and clustering of certain types of consumers. As some groups of consumers might exhibit a high price elasticity while others do not, the implementation of specifically tailored dynamic tariff schemes can help to ensure customer retention, while at the same time influencing aggregated demand in a beneficial way for the energy provider.

Type 7—*What-if scenarios*: A powerful functionality within the energy management system is a planning tool that allows to calculate the effects of adjustments or new installations in the grid. For instance, large amounts of additional PV panels, wind turbines, or battery storages might cause large effects on a certain part of the grid or even to the whole system. The tool allows estimating these effects prior to the installation. The connected benefits therefore arise indirectly through the transformation of existing business processes.

Type 8—*Investment decisions and root problem detection*: Lastly, the energy management platform can yield benefits that are only realized in the long run on an indirect and managerial basis from restructuring whole business processes. For instance, an active demand-side management reduces dependencies on suppliers of backup capacities. This can prove beneficial for the negotiation of future long-term contracts or the portfolio management at wholesale markets. Additionally, benefits arise from the detection of root problems in the infrastructure or existing business processes.

9.3 Business Benefits for Related Use Cases

This section contains a brief description of the use cases that are related to the CDSS on an operational level, as well as an outline of the associated benefits in the energy market.

9.3.1 Creation of City Energy View

A key component of smart grids and other energy supply systems is to have sufficient information about the current status of the system through a variety of measurement points. These embedded processing and digital communications enable the energy grid to be:

- observable (able to measure the states of all grid elements)
- controllable (able to affect the state of any grid element)
- automated (able to adapt and self-heal)
- fully integrated (fully interoperable with existing systems and have the capacity to incorporate a diversity of energy sources).

The ability to display all relevant information from the supply chain to the end consumers, most importantly supply and demand levels, is one of the central and basic components of the smartDSS. It will include data for historical, current, and forecasts for future time horizons:

- Visualization of the near real-time energy status of the city.
- Visualization of the forecast (24/48/72 hours) for the energy status of the city.
- Visualization of the history (e.g., last day/week/month/year) for the energy status of the city.

To capture and display, all relevant data can create already a variety of different benefits, such as the detection of inefficient consumption patterns. The smart meter data capture continuous consumption behavior in the buildings in the pilot cities with installed equipment. As was found out during the first period after the installation of smart meters, one school building in Rijeka was fully heated over weekends when building was vacant. It turned out that the housekeeper responsible for the building was leaving the heating on over the weekend. The analysis of smart meter data helps to detect those inefficiencies in consumption patterns and significantly reduce heating costs.

The city energy view has the option to display all relevant consumption and production data as basis for the decision making process. The identification of trends in the consumption and production data (seasonalities, working day/weekend, hours of the day) allows the effective management of the supply and demand side. For instance, the detection of certain trends can help to optimize the purchases of needed quantities of energy supply, e.g., at the wholesale market. This is because it is usually cheaper for an energy retailer to make purchases of needed quantities early in advance. Short-term energy supply purchases are mostly connected with higher risks, e.g., higher price volatility at the intraday market, and necessity to activate balancing energy. Alternatively, peaks in consumption can be met through the creation of new dynamic tariffs or specific demand response events. For this, it is needed to clearly detect when exactly peaks in demand occur and how strong they are. Equivalently, the connection of certain weather conditions to production and consumption levels is enabling the utility to make better forecasts, which in turn allows an effective supply and demand-side management.

Connected benefits to the creation of a city energy overview are therefore:

- Tool for quickly capturing an overview of data and energy flows.
- Detection of inefficiencies or unusual production/consumption levels within the regional area.
- Basis for decisions about certain actions (e.g., sending out demand response requests, shutting off certain energy production units).
- Identification of bottlenecks in the system or areas for grid enhancement.
- Identification of trends in the consumption and production data (seasonalities, working day/weekend, hours of the day).
- Decision basis for portfolio optimization at wholesale market, i.e., day-ahead or intraday markets.
- Decision basis for the creation of new demand-side management measures, e.g., creation of dynamic tariffs or demand response events.

9.3.1.1 Testing and validation in the pilot of Plovdiv

The energy view option is the first step: The utility company can take to an effective energy management. With a view to the greater digitalization of the energy market, close to real-time energy view is a pre-condition for building the future complex processes and relations.

Through the monitoring tool, the utility can obtain the information needed for the decision making process which aims at energy efficiency and reduction in costs for investment and maintenance and repairmen. All the measures and decisions which are taken are on the basis of the information for the consumption (resp. production)—historical, current, and future forecast. After decision and implementation of action, the energy view option again gives feedback if the action taken achieves the desired results or not.

Thus, as stated, the energy view option is the basis on which the future actions are planned (Chart 9.1).

The CDSS provided to EVN TP and EVN EP (the heat and distribution grid companies), the means for monitoring the status of the consumption of heat (electricity) energy of a building, region, or the whole city, and at the same time to track also the energy production. The software presents a unique opportunity for grid operators (heat or electricity) to have an overview on the energy flows—production and consumption in close to real time.

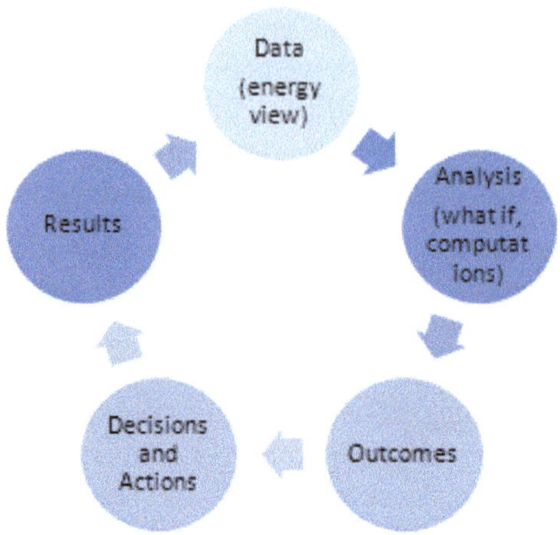

Chart 9.1 Data and decision process.

Thus, through the use of the CDSS, EVN TP has overcome a technical limitation in the use of a more user-orientated software which can be used both from technical and customer relations departments (Chart 9.2).

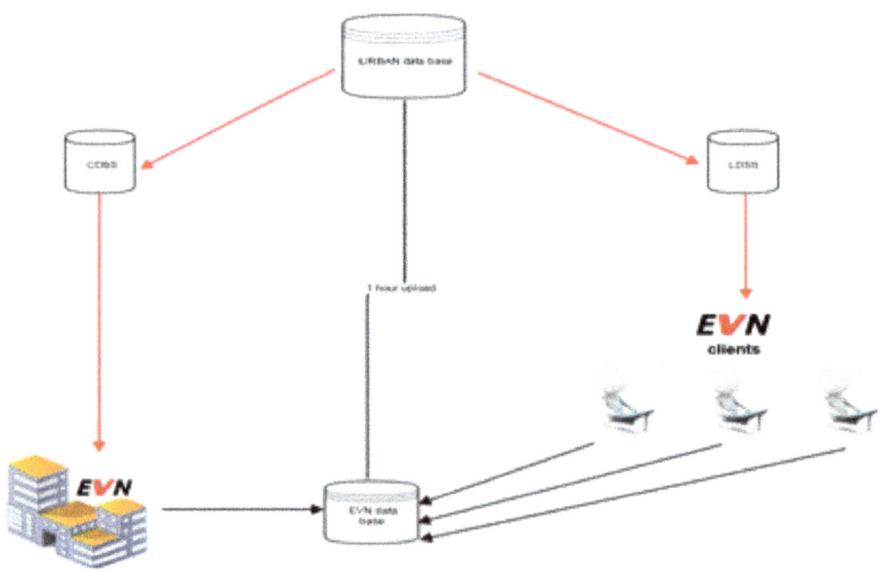

Chart 9.2 Use of the smartDSS in the pilot of Plovdiv.

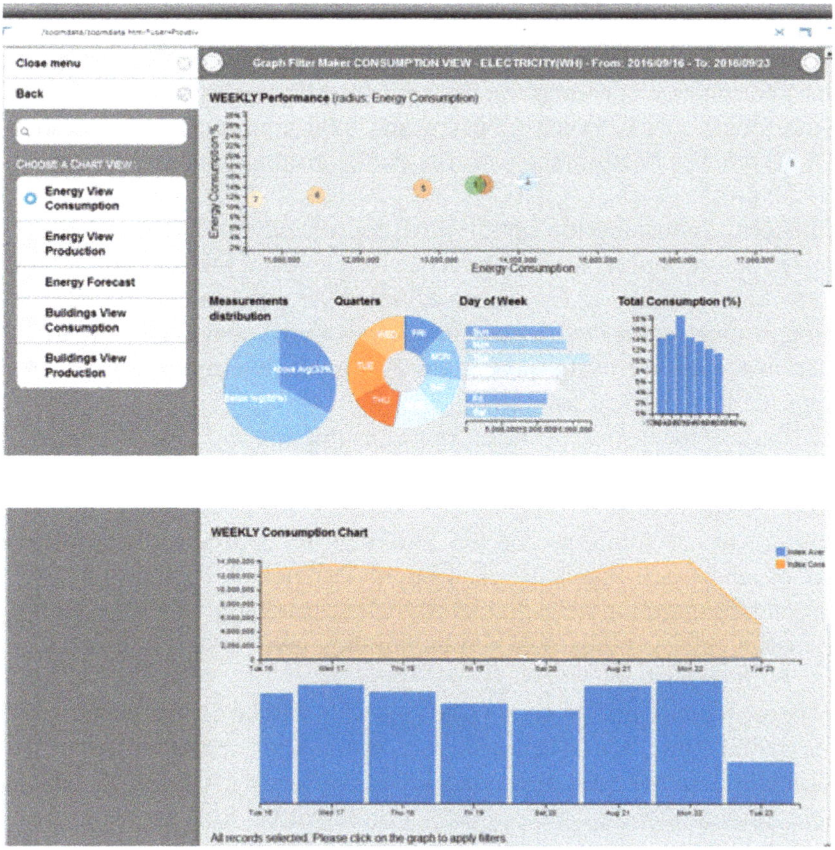

Figure 9.2 Plovdiv city energy weekly view.

The city energy view option is helpful in the energy planning and balancing. Thus, it is also supported by a forecast function. In this way, the actual use of the CDSS could help the utility companies to achieve higher reliability of the grid infrastructure and at the same time lower costs and achieve energy efficiency. If the software encompasses the whole city consumption, then the tool will be very helpful in terms of the medium- and long-term planning of investment in new capacities, facilities, and energy infrastructure as a whole. In the case of the city of Plovdiv which is period of re-industrialization and opening of new factories and production capacities, a good overview of the energy consumption and production equips the utility with a mechanism or monitoring and analyzing of the needs of a region or the whole city.

Through the energy consumption view option, we have monitored the consumption patterns in the participating buildings. We identified when the peak times for energy consumption are. For electricity consumption, these are the morning hours between 6:30 am and 8:00 am, and in the evening 18:00–21:00 pm. For heating, the peak hours of consumption are the evening 17:00–21:00 pm.

The historical consumption option provides a handy tool of tracking the deviations in consumption pattern which can be due to not just energy savings but some technical problems. For instance, Figure 9.3 gives an overview of the hot water consumption in Plovdiv and the days with the peaks. This provides an indication for the utility to have a closer observation and check for the reasons for the increased consumption.

On the basis of the historical and current consumption and pattern tracking, we could identify possible additional services which we can offer to the customers. When there is a potential for savings, we could offer or even develop tailor-based solutions for the buildings to optimize their energy consumption and reach higher level of energy efficiency. Furthermore, the building comparing option helps to identify whether the implemented energy efficient measures in a comparable similar building gives results compared to a building with no such measures taken (Figure 9.4).

The energy forecast option gives us an idea what will be the demand for the following period. Thus, this has allowed us as power plant operator to align the production based on the estimated consumption. However, the results which are visible through the software are only representative. Still the significant number of the DH customers in Plovdiv use metering devices which do not

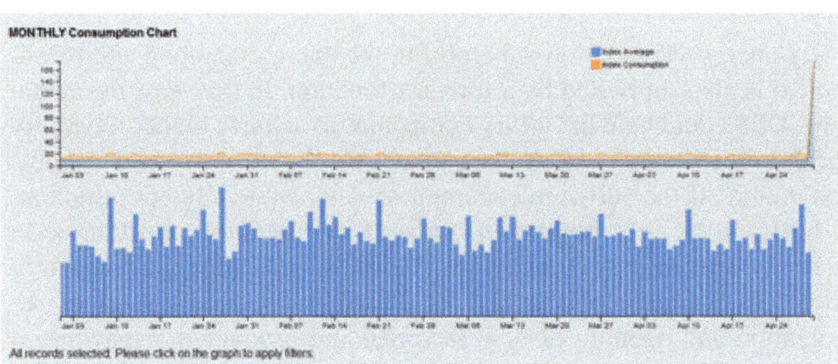

Figure 9.3 Hot water consumption in the city of Plovdiv (01.01.2016–30.04.2016, covering the heating season).

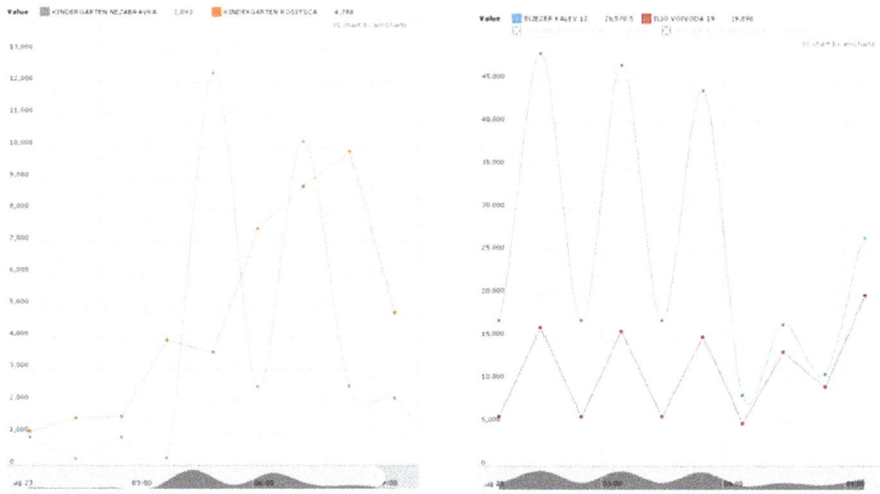

Figure 9.4 Electricity use over 1 day (23 August 2016)—comparison between two kindergartens (*left*) and two buildings with relatively similar characteristics (*right*).

allow hourly reading or their internal heating installation is vertical without individual metering devices.

9.3.1.2 Testing and validation in the pilot of Rijeka

The CDSS provided for municipal and utility users in Rijeka had the goal to provide users with consumption details for buildings included in project. Also, both types of users could make parallel testing and validation of data shown in the system with the data shown in other data systems (installed by manufacturer of metering equipment or other proprietary databases).

Municipal users had the opportunity to monitor consumptions of several energy sources used in each building and to help them decide what measures might be taken in the future to try and reduce (or control) the consumptions. From their experience with the CDSS, the municipality deems the following:

- Application is user-friendly
- It provides service management of energy consumption
- Users must have a basic technical knowledge to use this application
- It provides quick response with unusual consumption
- It reduces reaction time for detect failures and losses on installations or devices. That way, we can reduce negative impact on the environment and reduce the financial losses
- The user can control the consumption indicated in the invoice supplier.

The CDSS has enabled us to respond more quickly to energy issues. It enables to identify specific situations where there are discrepancies in consumption that were not consistent with historical data or operations. Being able to identify these anomalies, the CDSS allows to dig deeper to identify issues that could be corrected. Now, the municipality can respond quickly and avoid costs in the future as well as to control operations beyond its normal cycle.

For example, in one elementary school, we detected increased heat consumption over the weekend. In a short time, we determined the cause. We separated the heating system that is used by the school facilities (5 days a week) and the other parts of the building that are consuming heating energy the whole week. We could not say with certainty the total amount of savings achieved by this action as during the same period the school was reconstructed (new facade with 16 cm of thermal insulation).

Overall, the implementation experience of CDSS has been valuable. Increased visibility of energy issues has aided in goal setting and overall promotion of energy management efforts. CDSS helps us to do proactive energy management and building the City's energy program and metrics around it. Utility users had the same possibilities as municipal users but with the addition of some extra options that have been tested and reported through demand response program.

Energy view that has been tested here has shown great benefit to utility company because utility users could monitor consumptions and at the same time monitor the influence of user-driven actions on the gas grid that utility manages. As described in demand response paragraph, utility had tested part of interface that is used for communication with private users by sending notifications about how users could act in order to influence gas network and overall gas consumption.

Based on these data about user behavior and participation in given notifications and actions, utility would get feedback how to balance gas network and daily consumption curve. This type of system enables direct communication with customers and at the same time gives direct real-time readings that are needed in this type of energy business due to obligation of utility company to buy and reserve amounts of gas on everyday basis. This system, being quick and responsive, and paired with, e.g., SCADA system with gas prediction and optimization modules, gives great opportunities to utility companies to optimize their gas networks and costs regarding gas transport and overall management. If this software would be used to cover all aspects of energy consumption and all utilities, it would be excellent for energy planning and balancing actions, and therefore widely mentioned energy efficiency.

9.3.2 What-if Scenarios

The "what-if" analysis can be a powerful tool for the energy provider, DSO, TSO, or other market actors, such as municipalities, as it gives them the possibility to evaluate the impact of certain actions on the energy system.

Distribution grid operators, for instance, are responsible for the maintenance, expansion, and reinforcement of the low-voltage grid and at the forefront for integrating distributed renewable energy sources into the grid. While the traditional system required the delivery of electricity from large power plants to densely populated urban areas, distribution grids now have to cope with the feeding in of large quantities of renewable energies and thus are required to support bidirectional energy flows. Demand response, along with other measures, can serve as alternative to extensive grid capacity enhancement, as it is able to ease temporal congestion of the grid. Large amounts of small decentralized production units, such as solar panels and windmills, can oftentimes be a big challenge for securing the stability of the grid, as their production is intermittent and difficult to predict. Therefore, grid operators can use the "what-if" functionality of the smartDSS to simulate certain events. For instance, the feed in of large quantities of solar and wind power on a sunny and windy day or an outage of an important power plant.

Equivalently, "what-if" scenarios can be used to detect or evaluate the necessity and benefits of investments in new infrastructure. This could be the construction of a new array of solar panels with a large capacity in a remote area of the grid system. Simulations of how the influx of large quantities of renewable energies might influence the grid system can identify the need for additional investments in infrastructure and grid enhancement.

Energy retailers can optimize their portfolio of buying energy at the spot market if they can evaluate the impact of certain weather conditions on the overall levels of energy production and consumption. An execution of a "what-if" analysis for the future (e.g., 24, 48, 72 h) for the energy status of the city allows making precise decisions about energy transactions at the day-ahead or intraday market. For instance, when it is forecasted that the next day is going to be cloudy and windy, this results in certain production levels from solar and wind power and consequently in certain requirements for short-term portfolio optimization.

In general, the "what-if" analysis can create the following benefits:

- Short-term portfolio optimization at day-ahead or intraday markets for energy retailers in response to expected weather conditions

- Simulation of certain levels of solar and wind power production and their effects on the energy system
- Better possibilities to maintain grid stability for grid operators by the accurate activation or avoidance of balancing energy
- Possibility to identify most suitable production units for shut-off in case of overproduction of renewable energies
- Investment decisions in new capacities or grid enhancement when bottlenecks are detected
- Evaluation of effects of the installation of additional power plants, such as windmills or solar panels.

9.3.3 Auditing/Billing

The functionality of auditing and billing allows the energy provider to manage and implement new tariff schemes efficiently. This will be especially valuable as the number of available tariffs is expected to increase due to the liberalization of the market and the possibility to implement dynamic tariffs.

With the implementation of such tariffs, the energy sector can balance out demand levels and benefits from a reduced infrastructure needed to generate and distribute power at peak times. It can also cut energy procurement costs through lower peak prices and reduce vulnerability to service failures such as blackouts.

Responsive demand driven by dynamic pricing can also reduce greenhouse gases and local pollutants. Enhanced price signals can cause customers to shift demand away from peak times, avoiding emission-intensive generators used to serve system peak in some regions. Customers may also cut demand entirely due to enhanced price signals and better consumption information from smart metering.

- Fixed—in which the consumer is charged the same amount for the electricity used no matter what time of day it is used. A fixed rate energy tariff is typically set for 1–2 years. During that time the consumers pays the same amount for their electricity regardless of any price changes in the market.
- Time of use (TOU)—in which electricity prices are set for a specific time period in advance, typically not changing more often than twice a year. Prices paid for energy consumed during these periods are pre-established and known to consumers in advance, allowing them to vary their usage in response to such prices and manage their energy costs by shifting usage to a lower cost period or reducing their consumption overall.

- Critical peak pricing (CPP)—High price periods may or may not occur depending in the status of current consumption and production levels. Price signals are provided to the user on an advanced or forward basis, reflecting the utility's cost of generating and/or purchasing electricity at the wholesale level.
- Real-time pricing (RTP)—Prices may change on an hourly basis or even in short time periods. Price signals reflect the current status of the grid or wholesale market prices as well as the general level of production and consumption.

Customers can benefit from a dynamic tariff if their overall energy bills decrease. As the energy provider does not need to charge risk premiums on top of a single tariff, the overall price levels can decrease from the adoption of dynamic tariffs. In general, peak demand is one of the most expensive cost parts for the energy provider (e.g., keeping backup generators available). Therefore, with the ability to influence the demand side, the energy provider might be able to reduce overall costs, allowing it to forward parts of the savings to customers as reduced tariffs.

The functionality to manage tariffs at the smartDSS can have further benefits for the billing process for customers, as it possible to accurately measure when and where energy has been consumed. This can help to create detailed bills on an automated basis. Being able to take exact measurements of consumption without having to read out meters manually any time it is necessary, can reduce costs of these process significantly. It is easier for the energy supplier to detect which customers are consuming energy but not paying their bills. Complaints and law suits from customers might be settled easier and faster as it is possible to proof exact amounts of consumption on an automated basis.

9.3.4 Technical and Non-technical Losses

Generally, in electricity supply to final consumers, losses refer to the amounts of electricity injected into the transmission and distribution grids that are not paid for by users. Total losses have two components: technical and non-technical. Technical losses occur naturally and consist mainly of power dissipation in electricity system components such as transmission and distribution lines, transformers, and measurement systems. Non-technical losses are caused by actions external to the power system and consist primarily of electricity theft, non-payment by customers, and errors in accounting and record-keeping. These three categories of losses are respectively sometimes referred to as commercial, non-payment, and administrative losses.

From a regulatory or governmental perspective, it is beneficial to reduce the total amount of technical and non-technical losses as much as possible, as its reduction means that less energy has to be produced overall. This is generally connected with lower costs in the energy sector, which has benefits to the society overall.

Especially, non-technical losses represent also an avoidable financial loss for the utility. As an effect, customers being billed for accurately measured consumption and regularly paying their bills are subsidizing those users who do not pay for electricity consumption. A classic case is a theft of electricity through an illegal connection to the grid or manipulation of a consumption meter. But examples also include unmetered consumption by utility customers who are not accurately metered for a variety of reasons.

Detect inefficiencies in power distribution

Having a variety of different measurement points between the point of generation of energy, its transmission and distribution, and the final area of consumption helps to detect where any losses on its way occur. Especially for electricity, the supply chain typically involves the transfer of electricity through the high-voltage grid, medium-voltage grid, and the low-voltage grid. At the same time, several market actors are involved in the supply chain, including producers, TSOs, DSOs, and the energy retailer, which typically have incomplete information about the status in the other parts of the delivery chain.

Having measurement points at these different parts of the supply chain can help to detect at which part inefficiencies or defaults occur. The visualization on a graphical map makes it possible to track exact locations. The information taken from the smartDSS allows restructuring process or enables decisions about investments in the corresponding infrastructure.

Identification of energy theft: Energy theft is a serious problem in many energy markets worldwide and can occur in many different forms, the most prominent contain the following:

1. Direct hooking from line, where the consumer taps into a power line from a point ahead of the energy meter. This energy consumption is unmeasured and procured with or without switches.
2. Bypassing the energy meter, where the input terminal and output terminal of the energy meter is short-circuited, preventing the energy from registration in the energy meter.
3. Injecting foreign element into the energy meter, where meters are manipulated via a remote by installing a circuit inside the meter so that the meter

can be slowed down at any time. This kind of modification can evade external inspection attempts because the meter is always correct unless the remote is turned on.

4. Physical obstruction or other form of manipulation of the smart meter. Often a foreign material is placed inside the meter to obstruct the free movement of the disk with traditional meters. A slower rotating disk signals less energy consumption.

With the smartDSS, significant deviations between the smart meter data at building level and the sum of smart meters data at apartment level are detected. Alternatively, the smartDSS can also filter for regional or district level. This makes it much easier to identify any form of energy theft, as the supplier does not have to rely purely on the energy meter. In addition, with a rollout of smart meters, it is not directly possible to physically alter or manipulate the work process within the meter.

Identify non-payment by customers: The identification of non-payment by customers can especially valuable for the case of heating in big residential buildings with multiple parties that rely on heat-cost allocators. Due to the characteristics of physical distribution of heat through the residential buildings, it is not possible to completely exclude some customers from heating even though they choose not to pay their bills. If some customers decide to opt-out of their tariffs, the fixed part of the costs of heating supply gets reallocated to the remaining customers in the building, which in turn have increasing incentives to opt-out. Therefore, customers that do not share their part of the costs are getting subsized by the paying customers, as apartments still receive heating from the flow through the building. As a result, in energy suppliers, such as EVN in Plovdiv, face large numbers of law suits from customers who feel that they are unjustly treated, mostly those who have to take the additional costs.

Being able to take exact measurements of consumption without having to read out meters manually any time it is necessary, can reduce costs of these process significantly. It is easier for the energy supplier to detect which customers are consuming energy but not paying their bills. Complaints and law suits from customers might be settled easier and faster as it is possible to proof exact amounts of consumption on an automated basis.

Avoid errors in accounting: Similar as to the detection of energy theft, the smartDSS here has the functionality to detect significant deviations between the smart meter data at building level and the sum of smart meters data at apartment level. Therefore, the energy utility can double-check any errors that might have occurred during the metering, accounting, or billing processes. In addition, processes of re-billing, e.g., after a customer complaint, can be done much easier and faster, which is connected to lower costs of the billing system.

9.3.4.1 Testing and validation in the pilot of Plovdiv

The so-called technical losses in the heat distribution grid refer to the costs from transportation of the energy from the point of production to the point of consumption in Bulgaria. As with all other goods, for the good "heat energy," there are also costs for transportation of the good between the producer and the consumer. These costs in the transportation of the energy are not paid as energy by the customers as they cannot be avoided and are associated with the transportation of heat energy to the buildings. They are taken into consideration by the Regulator in Bulgaria when determining the end price for heat energy for the DH companies. Thus, they are not directly covered by the end customers but are part of the end—price for the heat energy.

Estimated production processing costs and realization of thermal energy for heating, cooling, and domestic hot water/hot water/for Plovdiv over the years will be as follows:

In contrast, what is denoted in the CDSS software as technical losses is actually the loss of energy (for heating and hot water) in the heating substation and the internal installations of the buildings as it measures the difference between the energy supplied to the building and the energy consumed in the individual apartments. However, this is not necessary a loss as such, as the internal installation also releases heat for which energy is needed and it cannot be considered a loss as the energy is used to warm the building same as with the radiators in the individual apartments.

In accordance with the Energy Act in Bulgaria, the internal installation for heating and for domestic hot water in a multi-dwelling building is considered also part of the "common parts" of the building, and hence, the maintenance and management of the common parts rests with all the owners in the buildings

Table 9.1 Estimated production processing costs and realization of thermal energy

Activity	Measure	2014	2015	2016	2017	2018	2019
Production	MWh	310,833	353,257	355,804	357,389	358,454	361,476
Technology expenses	MWh	118,948	132,057	128,737	126,675	123,756	121,474
Technology expenses	%	38.27	37.38	36.18	35.44	34.52	33.60
Realization	MWh	191,885	221,200	227,067	230,714	234,699	240,002

in proportion to their share percentage of the common parts. In the same vein, the costs for the energy used in the internal heating installation are shared among the apartments using the service in proportion to the heated volume of their apartments. The distribution of the heat released by the internal building installation is made following a complex methodology approved in Ordinance [11]16-334/2007 for share distribution of heat energy in multi-dwelling buildings ("the methodology"). The methodology incorporates a complex mechanism which takes into account the type and the thermo-physical characteristics of the building and the heating system.

However, the CDSS took a more generalized approach which can be applied in a number of different situations and does not follow strictly the approach set out in the methodology in Bulgaria. Thus, the approach of the CDSS can be applied to various buildings in various countries and the logic stays the same—difference between supplied energy to the whole building and sum of the consumed energy in the individual apartments (Figure 9.5).

The function in the CDSS gives a basis for review of the mode of consumption for different buildings and provides a basis for analysis why the "technical losses" in certain buildings are lower than in others and what measures are taken to diminish them. This analysis is a basis for the utility to make a plan what kind of services for maintenance and improvement can be offered to a building with high values of technical losses. Thus, the CDSS offers the possibility for creating building profile and possible list of consultancy or additional services which can be offered to customers in this building.

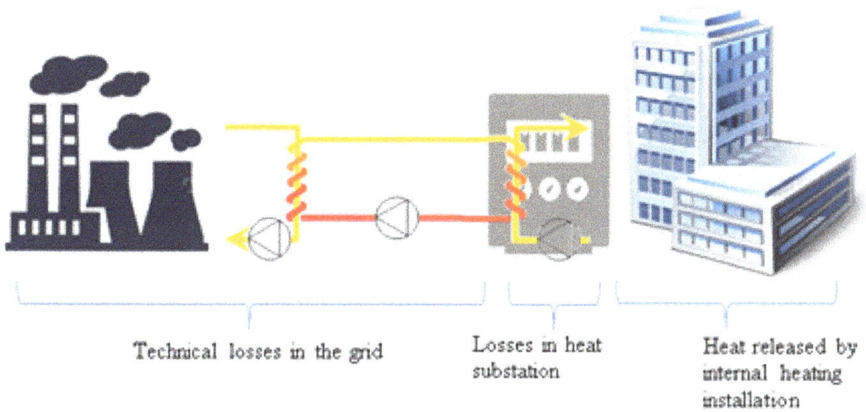

Figure 9.5 District heating grid scheme.

Furthermore, the function in the CDSS is also a helpful tool to detect some deviations in the technical losses and investigate the reasons for. They can be due to leakages, improper use of energy, or some technical problem. At least in Bulgaria, the maintenance of the internal installation is a responsibility of the inhabitants of the building, but the information for the technical losses in the CDSS represents an overview of the building and gives signals for possible improper use. Overall, reduction in the technical losses leads to reduction in the consumption of energy and less GHG emissions.

9.3.5 Demand Response

9.3.5.1 The model

Demand response relates only to electricity energy. Through demand response, final consumers (households or businesses) provide flexibility to the electricity system by voluntarily changing their usual electricity consumption in reaction to price signals or to specific requests, while at the same time benefiting from doing so.[1]

In order to provide the necessary IT tools to make this happen, the CDSS provides the following functionalities:

- User can create/modify/delete DR programs (every DR program is associated with a peak of energy consumption);
- User can get information about peaks of energy consumption;
- User can manually create a peak of energy consumption;
- User can create/modify/delete notifications related to associated peak of energy consumption;
- User can get information about peak of energy consumption;
- User can get information about the status result about every notification associated to a chosen peak of energy consumption.

Testing and validation in the pilot of Plovdiv

The demand response business case was not implemented in the city of Plovdiv. There are two major barriers before the implementation of demand response in Bulgaria at the time the project was realized:

- Regulated prices by law for customers connected to low-voltage network and no justification of expenses for rewards for shift in the consumption.

[1]Eurelectric, "Everything You Always Wanted to Know about Demand Response," Dépôt legal D/2015/12.105/11, available at <http://www.eurelectric.org/media/176935/demand-response-brochure-11-05-final-lr-2015-2501-0002-01-e.pdf>, p. 3 ("Eurelectric Demand Response Report").

- No economic benefit for customers to switch to the free market due to regulation of the prices for purchasing electricity and margin of the utility.

9.3.5.2 Regulatory environment

As it has been mentioned on several occasions in previous documents, demand response has not been tested in Plovdiv due to some legal impediments which prevented implementation.

The full liberalization of the electricity market in Bulgaria was launched as of July 1, 2007 and it is a step-by-step process which started with liberalization of the wholesale market and gradually continues with the retail one.

Customers connected to the low-voltage (residential and small business customers, i.e., the participants in iURBAN project) are still on the regulated market where the prices are regulated and fixed by the Energy and Water Regulation Commission of the Republic of Bulgaria ("EWRC").[2] The prices determined by the Regulator are not a ceiling or a framework within which the prices should be. On the contrary, the prices approved by the EWRC are fixed and binding on the DSO and the electricity supplier and represents the price which is billed to the end customer.[3] The EWRC stipulates the grid connection and distribution fees, the price for electricity for the end customer (BGN per kWh), and other ancillary services for a certain regulatory price period (which is usually a year and starts on 1st of July and continues up until 30th June in the following year).[4] All customers participating in the project are supplied with electricity on regulated prices by the company EVN Bulgaria Elektrosnabdiavane EAD.

Furthermore, it is essential for demand response to provide a reward to induce the shift in the consumption. The current situation of the regulation in Bulgaria does not allow expenses for rewards for shift in the consumption to be justified under the regulated market prices. From accounting and tax perspective, the company would make an expense and next to it there is no profit or benefit because the EVN Bulgaria Elektrosnabdiavane EAD buys electricity on flat rate and sells it to customers on fixed price. Such actions on the regulated market could certainly raise concerns and check by the authorities.

---

[2]See Article 21 and 30 (1) of the Energy Act.

[3]See Ordinance No. 1 of 18.03.2013 on the Regulation of Prices for Electricity issued by the Chairman of the State Energy and Water Regulation Commission ("Ordinance on the Regulation of the Prices for Electricity").

[4]See Article 30 (1) of the Energy Act.

9.3.5.3 No real economic benefit

Customers connected to the low-voltage can opt out and choose a supplier of electricity energy and negotiate the price for electricity with it (important note is that the grid service fees are still regulated). However, customers do not have economic incentives to go on the free market, as currently the price determined by EWRC for household customers is significantly lower than what is offered at the liberalized market due to the cross-subsidy of the prices between business and household on the regulated market. EWRC artificially keeps the prices of the electricity for the household customers low at the expense of the prices for the business customers. Thus, there is no real price competition on the market and the household customers (mainly the participants in iURBAN) have no economic benefit to switch to the free market where the supplier can offer more flexible price options and include demand response as well.

To illustrate the situation, the average price for electricity at the regulated market currently is 106 BGN/MWh or 69,50 BGN without the public obligations fee. The price for electricity at the liberalized market is at the average—76 BGN/MWh. The difference of ab. 6, 5 BGN/MWh is obvious and enough deterrent for residential customers to switch to a new supplier.

Testing and validation in the pilot of Rijeka

Demand response program in Rijeka has been tested and implemented with 8 private users. Selected customers use natural gas as a primary heating energy source. The largest gas consumptions occur in the morning (period before people go to work) and in the afternoon (when people are back from work).

Misbalance in consumptions during day is not good for the distribution network, so the goal of DR actions was to influence consumption peaks in gas optimization. Users are motivated to set their thermostats in accordance with the requirements of the DR action in order to reduce the consumptions during the periods of increased consumptions. Demands are appointed by the utility in relation to the consumption overshoot.

At first, enControl needed to be installed. It is a system for home management and control of energy consumption. Through this system, it is possible to operate the home heating system in such way that the temperature of each radiator (thermo-head) and room thermostat can be set separately to achieve most of level comfort for users, while at the same time, users can try to be most energy efficient (lower temperature set points for rooms that are not that frequently used, etc.). The control is provided both through mobile application (smartphones) and online via cloud service application where each user can log in with assigned username and password.

DR actions have been created a few days before a certain optimization needed to take place and were sent to users in the form of a notification through the Web application. These notifications could then be forwarded to users via any of pre-set ways—via e-mail, SMS, or mobile application (according to the user's selection). The tasks users needed to do in order to fulfill the request consisted of some form of action regarding the main thermostat. Users were instructed the temperature that needed to be set and the time period for which the setting needed to be active (e.g., set your thermostat to 21degC or below, time period: 7 AM–9 AM). If the user would successfully complete the request, he would be notified about successful participation in DR action and would be awarded a coin. Additional coin was assigned to users who fulfilled at least five DR actions during each month of the program. Users who have met all the required DR actions would be awarded additional five coins.

Table 9.2 gives an overview of all DR actions and results for each user. Every DR action shown in the table has several additional pieces of information like the time when notification was sent (scheduled time) and the time when the action took place (action time). Furthermore, it shows the required thermostat temperature during the DR action and outside temperature during this period. In part of the table which gives an overview of the performance for each action by individual user, notification delivery report (first column) is shown. The second column shows whether the user has successfully performed a given task. The aim of each particular DR action is given at the bottom of the table, as well as additional remarks related to the action.

During the demand response program, 19 DR actions took place. The first two were unsuccessful because of notification sending and receiving issues, but the following 17 DR actions were conducted successfully. Altogether, 132 demands were sent by utility and received by users which resulted in 96 demands that have been completed, in total 72.7% positive responses.

It is important to note that participants of DR programs can be described as younger and middle-aged users, familiar with everyday use of computers and other IT devices, so general usage of Web (and mobile) application did not represent a problem for them. Despite all mentioned, the large level of engagement needed for DR actions followed by high rate of sent DR notifications finally did represent a problem for almost all users and had become too intense for them.

Table 9.2 Executed demand response events number 15 to 19

	DR15		DR16		DR17		DR18		DR19	
Installations	Below Avg. Temp. during DR	20	Below Avg. Temp. during DR	20	Below Avg. Temp. during DR	19	Below Avg. Temp. during DR	18	Below Avg. Temp. during DR	17
		14,11		12,96		12,29		11,9		15,42
	Scheduled time (UTC)	Action time (UTC)	Scheduled time (UTC)	Action time (UTC)	Scheduled time (UTC)	Action time (UTC)	Scheduled time (UTC)	Action time (UTC)	Scheduled time (UTC)	Action time (UTC)
	31.03.2016 10:00	01.04.2016 17:00–19:00	01.04.2016 12:00	03.04.2016 14:00–16:00	05.04.2016 09:00	06.04.2016 05:00–07:00	08.04.2016 09:00	09.04.2016 14:00–16:00	11.04.2016 11:00	12.04.2016 06:00–07:00
Pvt1										
Pvt2										
Pvt3										
Pvt4										
Pvt5										
Pvt6										
Pvt7										
Pvt8										
Remarks	Summer time—UTC+2h	Summer time—UTC+2h	Summer time—UTC+2h	Summer time—UTC+2h	Summer time—UTC+2h	Summer time—UTC+2h	Summer time—UTC+2h	Summer time—UTC+2h	Summer time—UTC+2h	Summer time—UTC+2h
Purpose of DR		To flatten the consumption peak		To see how much they are willing to reduce their comfort while at home		To flatten the consumption peak		To flatten the consumption peak		To flatten the consumption peak. The lowest required temperature.

As conclusion, if results shown here could be achieved in real terms, on a daily basis, flattening of consumption peaks would be achieved. Therefore, the purpose of the DR program would be accomplished.

9.3.5.4 Demand response—lessons learnt

- The utility is performing DR program for peak shaving on its supply side. Expected savings should be used to cover incentives to the customers who successfully fulfill DR condition and to partially cover the investment in smart installation needed to pursue DR actions. Customers shift their consumption toward before or after the DR interval. Utility's intention is not to reduce the consumption in general, but to shift it from the peak interval. If the overall consumption decreases a lot, the whole DR concept might be endangered.
- Installation in 8 private homes consists of the following: thermostatic valves, smart thermostat, smart plugs, magnetic door/window sensors, motion detector, thermometer, humidity meter, lux meter, and gas smart meter. Each system measured overall gas consumption and partial electricity consumption (smart plugs).
- Initial installation costs for smart homes and smart installations used in the project are very high to justify its feasibility, in iURBAN that was done on purpose, to test and prove usability of different components. In further cases, if only DR is objective, the installation cost should be kept low—just GW and remote thermostat. Thus, utility can co-finance that investment from future savings based on DR actions. If customers want more functionality, the rest of installation could be bought by them.
- Customers reacted very well, but probably also due to the fact that they knew they were part of a test. It is expected that, in the case of the long-term commitment, the level of successful DRs would be lower.
- The customer's feedback related to complexity of the DR process shows that the level of automation needs to be increased, especially on the customer's side, in order to make a viable use case => less interaction from customer needed. If the level of automation can be increased, we also expect that the needed financial incentives can be lower, as there is less discomfort for the customers to deal with the DR request.
- Customers have expressed their wishes to not interact, but they want to keep option to "overwrite" DR requests, i.e., to keep final control over consumption. On the other side, that would reduce savings on utility's side and lower the DR incentives. One further research question could

be directed toward exploring how often customers use the overwriting function.

- Number of customers included in the DR program was 8. The sample is quite small; the program needs to be tested on a larger scale to give more relevant results.
- Business case does not seem to be very favorable for gas, but maybe more for electricity. It strongly depends on saving potential, specific for each market/country. The most feasible gas DR program is viable only for customers who use gas for heating/cooling, where the consumption is significant enough and that leads to sufficient savings for utilities to finance investment and incentive costs.

9.3.6 Variable Tariff Simulation

9.3.6.1 The model

Variable tariffs are the basis of the implicit demand response (also called "price-based")[5]—consumers are responding to time-varying electricity prices that reflect the value and cost of electricity in different time periods. Consumers can shift their electricity consumption away from times of high prices and thereby reduce their energy bill. The economic benefits are duly described in Deliverable 6.2.

The CDSS software is also designed to allow the energy supplier to offer to its customers time-varying prices: Different tariff plans can be made through the CDSS and the supply can create/modify/delete tariff plans and can specify various attributes, for every tariff plan, like

- City: the city name where the tariff apply
- Tariff Name: the name of the tariff plan.
- Group: customer group to whom the tariff plan should be applied.
- Commodity.
- Unit: measure unit of commodity.
- Currency: The currency applied.
- Date/time period.

The software also allows for comparison of tariffs, getting information about tariffs costs (month by month) related to the last year.

[5]Eurelectric Demand Response Report, p. 3; ACER, "Demand Side Flexibility: The Potential Benefits and State of Play in the European Union", ACER/OP/DIR/08/2013/LOT 2/RFS 02, available at <http://www.acer.europa.eu/official_documents/Acts_of_the_Agency/References/DSF_Final_Report.pdf>, p. 5.

9.3.6.2 Testing and validation in the pilot of Plovdiv

As with demand response, the implementation of time-varying tariffs is not an option currently in Bulgaria for the customers participating in iURBAN project, although they have the technical equipment (i.e., metering devices taking data at 15-min intervals) to be able to participate in such a program.

The EWRC approved the tariff time zones in the seasons according to which different tariffs apply—day and night tariff for the household customers, and day, night, and peak tariff for the non-household once.[6]

Tariff zones for **non-household** customers:

November–March		April–October	
Day	6.00–8.00	Day	7.00–8.00
	11.00–18.00		12.00–20.00
	21.00–22.00		22.00–23.00
Night	22.00–6.00	Night	23.00–7.00
Peak	8.00–11.00	Peak	8.00–12.00
	18.00–21.00		20.00–22.00

Tariff zones for **household** customers:

November–March		April–October	
Day	6.00–22.00	Day	7.00–23.00
Night	22.00–6.00	Night	23.00–7.00

Thus, variable tariffs currently cannot be tested in Plovdiv since there are tariff zones (day and night tariff) approved by EWRC. Furthermore, in accordance with the Energy Act, the EWRC sets mandatory quotas for certain types of producers on the basis of which the public supplier—NatsionalnaElektricheskaKompania EAD (NEK)—sells electricity on the regulated price to the end supplier (supplying with electricity customers on the regulated market).[7] The energy suppliers buy electricity on a flat rate from NEK which is regulated by the EWRC.[8] EVN end supplier cannot sell electricity to customers at a price below costs because this qualifies as dumping and the company will be sanctioned by the competition protection commission.

[6]Price Decision Ц-002 of the State Energy and Water Regulation Commission as of 29.03.2002.

[7]See Article 21, para 1, point 21 of the Energy Act in connection with Ordinance on the Regulation of the Prices for Electricity.

[8]Article 15, para.2, point 2, of Rules for Trading with Electricity issued by the Energy and Water Regulation Commission, in force as of May 9, 2014.

Apart from the fixed time zones when tariffs apply, there is another limitation on the implementation of variable tariffs—the metering device in place of the customer. Although the customers participating in iURBAN project have metering devices allowing remote and hourly metering, not all customers in Bulgaria have such meters installed. On the liberalized market, customers which do not have the technical facilities for hourly metering have standard load profiles. Thus, variable tariffs actually cannot be applied as there are standard consumption curve applicable for the month. Furthermore, the Energy Regulator also in its decision from 2013 ruled that installation of smart meters is not mandatory and it is up to the grid companies to decide whether they would invest money in that.

In conclusion, there are two limitations on the implementation of variable tariffs in their strict sense and in the forms as described in Deliverable 6.2—regulatory and technical. As the different types of variable tariffs require frequent data gathering, this is a test for the remote metering system and communication. In a more simplified form, we can actually say that the day and night tariffs are a form of variable tariffs and as such have been tested within the iURBAN project. As the customers participating in the project, pay their energy according to the prices determined for the two tariffs by EWRC.

9.3.7 Consultancy Services

Lastly, to offer consultancy services to the end consumers can be an applicable business service for municipalities or third-party operators. Especially, municipalities or other local or regional authorities or organizations have typically a number of big buildings with many employees or visitors, such as schools, kindergartens, public libraries, and city council. The number of people that are frequenting these buildings is usually quite high, but these people do not pay for the costs of energy consumption, which increases the risk of inefficient consumption behavior. Therefore, if the municipality is operating the smartDSS, it can critically analyze the consumption patterns within the buildings and detect any sources of inefficiencies (e.g., heating or lighting at night or during weekends) or high consumption devices that might be turned off or replaced. Furthermore, optimal heating patterns can be established through smart heating devices or specifically educated energy managers for the building can be trained to keep inefficient consumption low. Necessary investments in better insulation within the building can be detected as this is oftentimes a major source of energy inefficiency, especially in countries such as Bulgaria and Croatia. In general, these services can raise awareness about consumption patterns and might positively influence behavior as a consequence. Private customers might benefit especially from these services as they reduce their overall energy costs.

If the characteristics of consumption exhibit certain peaks throughout the day/week or any high-load devices are existing, e.g., heat pumps or air conditioning, it might be financially profitable to engage in certain demand-side management measures. This could be

- Participation in specific demand response programs that are tailored toward the characteristics and needs of the building and consumers.
- Switching from fixed tariff rate to dynamic tariff if a high price responsiveness can be obtained, e.g., through the installation of smart devices that are able to shift/store loads or the appointment of an energy manager for the building.

Offering consultancy services is also a possible business service for third-party operators, as these can specialize in connected business models and build up further competencies. It is less applicable for utilities, specifically energy retailers, as their business model is based on selling as much energy as possible.

Potential benefits for the market actors can therefore be

- Possibility to make decisions about investment in better insulation within the building.
- Raising awareness about high consumption devices within the building or other sources of additional energy consumption, e.g., open windows.
- Detection of unnecessary consumption devices, e.g., lighting at night, air condition.
- Establishment of optimal heating patterns within the building, e.g., turning off heating at night or during times of vacancy.
- Decisions about investment in further smart systems, such as smart heating installations, automated air conditioning.
- Evaluation of profitability to take part in specific demand response programs, e.g., direct load control.
- Switching from fixed to dynamic tariff if high price responsiveness can be obtained, e.g., through smart devices or appointment of energy manager.

9.4 Conclusion and Policy Implications

In this chapter, we have outlined the outcomes from the testing and simulation of the selected use cases pertaining to general business case of the iURBAN project. Following the review made, it could be concluded that the use of

CDSS assists the daily business of utilities, municipalities, and probably also energy managers and creates new opportunities for their business models in the sense of providing the ground on which new services can be built (e.g., city energy view, technical loss) or providing opportunities for cost reduction (e.g., demand response), or allowing grid expansion without investment in new infrastructure (VPP and what if analysis). The analysis made for the European replicability also supports that and outlines the current situation with market enablers and obstacles for major European markets.

Consequently, the utilities will need to adapt their business models to capture the opportunities created for them by the new technologies and respond to the demands of their customers, which are becoming more active players in the new energy systems. In turn, municipalities will also need to amend their city policies to be in line with their citizens' needs or expectations. Energy and facility managers also compete on a rough market and offering services with added value such as smart energy management will give them competitive advantage. Thus, ICT tools such as CDSS comes as a right ally to help its end users to respond more adequately to the needs of their customers/citizens and position better on the market.

Data-driven technologies such CDSS will play important role in the energy systems, which are now undergoing transformation and become more and more digitalized. However, it becomes evident in the course of the project implementation, that certain use cases (such as demand response) could not yet be implemented in the pilot cities due to the regulatory and legal restrictions.

Therefore, the penetration of new technologies and the development of a sector such as the energy should go hand in hand with the reformation of the legislation and sufficient communication to the end users. Besides that, adequate protection of personal data and sensitive information (such as consumption pattern or way of bill payment) should be ensured. Through the implementation of use cases in the pilot, it has been highlighted that creating general understanding and knowledge in customers about innovative technologies is equally important for their acceptance and actual implementation as it had been the case with the pilot of Plovdiv (city energy view) and Rijeka (demand response). Finally, a European market for energy services can be created if there is a level of harmonization of national laws is achieved allowing the energy systems to function in similar market-based way and safeguarding the personal data of end users.

The implementation of the business cases of the CDSS calls for harmonization across the EU and corresponding changes in the legislation in different Member States, allowing the implementation of new technologies along with the demand response and dynamic tariffs which helps customers save energy, money, and GHG emissions.

The following general needs regarding adjustments in the regulatory framework have been identified in the pilot cities:

- Allow the flexibilization of tariff schemes, i.e., the implementation of dynamic tariff for end consumers and private households. This can both help to prevent critical load situations and the implementation of new innovative energy services. So far, grid fees and most other energy price components remain fixed, which hinders forwarding incentives down the value chain. The price regulation for customers on low voltage should be abandoned together with the firm fixation of the tariff plans.
- The participation of third parties in the energy sector is mostly not possible yet, e.g., for demand response aggregators or other service providers. Opening the market for new entrants should facilitate further innovative energy services, which in turn can increase the awareness of consumers about energy efficient behavior.
- It is recommended to enable regulatory innovation zones or settings that invite for experimenting with ICT measures and new business models. Investments into innovative smart energy tools are hindered by a high degree of uncertainty regarding the regulatory framework as well as a general lack of transparency with the decisions of the regulator in the energy sector.
- Existing balancing options are focused solely on generation. The integration of demand response measures can constitute a cost-efficient alternative. The inclusion of DR at the electricity spot market and ancillary service market is recommended to allow for effective business cases related to demand response. At the same time, pooling of different demand resources should be made possible in order to reach the necessary threshold and size regarding the minimum requirements at the different energy markets if prequalification for balancing energy can be made at the aggregated pool level.
- The communication between a presumably increasing number of market actors and with a large number of consumers should be made easier regarding the communication infrastructure. Therefore, we recommend the consideration to adopt communication standards for smart infrastructure from an early point on, e.g., the open ADR standard to standardize demand response events between different market actors with heterogeneous smart grid infrastructure. In general, technical communication

standards for the IT infrastructure could significantly reduce transaction costs for involved market actors and facilitate connected investments. Policy makers might encourage the use of such open standards by incorporating them in their digital agenda.

References

[1] Scholz, R., Beckmann, M., Pieper, C., Muster, M., and Weber, R. (2014). Considerations on providing the energy needs using exclusively renewable sources. Energiewende in Germany. *Renew. Sust. Energ. Rev.*, 35, 109–125.

[2] "Cities, Towns and Renewable Energy: Yes in My Front Yard", International Energy Agents (IEA), 2009.

[3] Boer, P. F. (1999). *The valuation of technology: Business and financial issues in R&*D. Wiley, New York.

[4] Curtin, J. P., Gaffney, R. L., and Riggins, F. J. (2009). The RFID e-valuation framework—determining the business value from radio frequency identification. In *Proceedings of the 42nd Hawaii International Conference on System Sciences.*

[5] Lee, H., and Özer, Ö. (2007). Unlocking the value of RFID. *Prod. Oper. Manag.,* 16(1), 40–64.

[6] Laubacher R., Kothari, S. P., Malone T. W., and Subirana, B. (2006). What is RFID worth to your company? MIT Sloan School Paper #224, Cambridge, USA.

[7] Baars, H., Gille, D., and Strüker, J. (2009). Evaluation of RFID applications for logistics: A framework for identifying, forecasting and assessing benefits. *Eur. J. Inform. Syst.,* 1–14.

[8] Feuerriegel, S., and Neumann, D. (2014). Measuring the financial impact of demand response for electricity retailers. *Energy Policy,* 65, 359–368.

Index

About the Editors

Dr. Narcis Avellana, PhD, CEO. Narcis Avellana is responsible for the vision, strategy, and leadership of Sensing & Control Systems. He is also responsible for the company's financial management. Under his leadership, S&C has grown to become one of the top 20 most innovative companies in Barcelona. Narcis co-founded S&C with Alberto Fernandez in 2006. Prior to S&C, he was a Marketing Director and Key Account Manager for the Application Specific Integrated Circuits (ASIC) Division at Epson Europe Electronics where he increased turnover from 200K€ to 7.5 M€. In 2001, he returned to Barcelona to establish the Seiko-Epson Business & R&D Center, where he was General Manager for 6 years. Along his professional career, Narcis has published more than 40 scientific articles and is an *Approved Expert Evaluator and Reviewer for* the European Commission. He received his Ph.D. in Microelectronics, *summa cum laude*, developed in the University of Ulm, Germany, his BA in Computer Science from the University of Barcelona and his MSc in Business Administration from the Autonomous University of Barcelona.

Alberto Fernandez is Operations Director in S&C for 10 years where he is responsible for developing hardware, software and firmware solutions. He has a track record of more than 20 years' experience developing technological solutions. Alberto holds a Computer Science MSc in microelectronics from the Autonomous University of Barcelona. He worked in the R&D department of the Spanish Microelectronics Centre for 3 years. Then he worked as R&D engineer at Microson, a Spanish leading company in the development of hearing aid products for the global market for more than 5 years. Following this, he was the Technical Manager of the Seiko Epson Barcelona Design Centre, at the R&D department for 6 years.

About the Authors

Aidan Melia is a project manager working on commercial and research and development (R&D) projects. Working for IES for over 3 years Aidan has managed 3 European projects, such as FP7 and H2020. His focus is on the areas of smart cities and ICT developments, as well as more recently on areas such as gamification. He has studied in a number of countries such as Ireland, the US, Spain and Germany.

Energy Agency of Plovdiv (EAP), Bulgaria is the first energy management agency established in Bulgaria under the SAVE II program of the European Commission (EC). The EC and others have considered EAP a huge success in the country and in the region since its inception in 2000. It is now a leading body in the field of sustainable energy development. EAP promotes energy efficiency (EE) and renewable energy sources (RES) on local, regional and national level, develops action plans, and performs feasibility studies promoting sustainable energy development. EAP also develops energy concepts and projects for municipalities and for small and medium-sized enterprises (SMEs), arranges financing, and provides expertise and consultations. EAP builds strategic partnerships with municipalities, governmental and non-governmental organisations, associations, networks, universities and businesses in Bulgaria and all other EU member states. It is a member of FEDARENE and ENER.

Fabrizio Lorenna received the High School Diploma in telecommunication specialization in 1986 and a Certificate of Professional Qualification for Computer Technologist, at ELIS Centre (Rome), in 1992. Since November 1992 he works at the company Vitrociset S.p.A. (Rome) as Analyst Programmer. During his work, he was employed in various projects mainly regarding: Satellite data management (production chains, archiving, data quality control, image processing) at ESA-ESRIN (European Space Agency—European Space Research Institute) in Frascati (Italy); Middleware, following CORBA standard, used for ATC (Air Traffic Control) at Selex-SI (Finmeccanica) in

Rome; "Billing Request" for the department Business Information Systems of Vitrociset S.p.A.; Tool for the management of an hardware inventory regarding the Lazio Region, for LAIT S.p.A. in Rome; Development and Research project for "Reward" and "iURBAN".

Jens Strüker has a habilitation in business informatics and economics. He is dean in the Faculty of Economics and Media at Fresenius University of Applied Sciences in Frankfurt and managing director of the Institute of Energy Economics (INEWI). His research focus is the identification, evaluation and marketing of flexibility in energy generation and consumption.

Karwe Markus Alexander is a doctoral candidate at the University of Freiburg, Germany. He obtained his Diploma with honors in Information Science at University Siegen. His research is in the field of Smart Metering Privacy with special focus on Privacy Enhancing Technologies in Residential Demand Response Systems.

Marco Forin is a computer science engineer with more than 10 years' experience in developing of technological solutions. Working in Vitrociset, he has the technical leadership of commercial and R&D projects, such as FP7 and H2020. During the last years, Marco has got to work on different applicative domains, starting from real-time systems for Italian Defence, to Smart City solutions for public administrations. Marco received a Bachelor Degree in Computer Science Engineering at "La Sapienza" University of Rome.

Mike Oates PhD, is technical analyst on commercial, and research and development (R&D) projects. Working for IES for over 3 years Michael has been technical lead/analyst on 6 European projects, FP7, H2020, and Marie Curie etc., http://www.iesve.com/research. Research project topic areas included glazing (electrochromic glazing), retrofit technologies, manufacturing, city modelling, and application development including gamification.

Sergio Jurado has 5 years of experience in the energy sector and is specialised in electric systems. He is a Ph.D. candidate in the Universitat Politecnica de Catalunya (UPC) in Artificial Intelligence applied to the energy efficiency. He worked in areas such as quality control, production line and teaching computer science in C++ for two years. He also worked in the R&D department of S&C for FP7, EUROSTARS and other R&D European projects as project engineer, project manager, as well as, performing data analysis tasks, developing and integrating Artificial Intelligence algorithms for the energy domain, and in the systems developed by the company. His Ph.D. is based on improving the predictions of consumptions and productions for prosumers and how negotiations can be performed among key stakeholders in the future Smart Grid. Nowadays, he is business manager of energy and utilities in a technological consultancy firm.

Sofia Aivalioti works for the R&D department of S&C as a dissemination and exploitation officer for European projects. She has a 5-years' experience as a consultant for international and regional development projects with specific focus on environmental sustainability. She worked for the Ministry of Finance of Greece for the finalization of the cross-border cooperation programme Interreg III-A and on FP7 and H2020 R&D project. She received a B. Eng in Civil Infrastructure Engineering from the A.T.E.I. of Thessaloniki, a M.Sc. in Environment and International Development at the University of Edinburgh and the Joint European Master in Environmental Studies – Cities & Sustainability where she also worked on a research project for the Columbia Law School.

Stefan Reichert is a doctoral candidate at the University of Freiburg, Germany. He obtained his M.Sc. in Economics at University of Freiburg and his B.Sc. at University of Heidelberg. His research focuses on the evaluation of flexibility in energy generation, consumption and storage, as well as data-driven business models in the energy sector.

For Product Safety Concerns and Information please contact our EU
representative GPSR@taylorandfrancis.com
Taylor & Francis Verlag GmbH, Kaufingerstraße 24, 80331 München, Germany

www.ingramcontent.com/pod-product-compliance
Ingram Content Group UK Ltd.
Pitfield, Milton Keynes, MK11 3LW, UK
UKHW020156260926
475702UK00047B/296

* 9 7 8 8 7 9 3 5 1 9 1 0 7 *